U0172555

Kees Kaan
卡斯·卡恩

Nathalie de Vries
娜莎莉·德·弗里斯

Jacob van Rijs
雅各布·凡·里斯

Francine Houben
法兰馨·胡本

Kas Oosterhuis
卡斯·欧斯特豪

Rethinkir

Mode

Dutc

Architectur

Dialogu

wi

Contempora

Dutc

Architec

国家自然基金资助，编号：51378210

再认识

荷兰现代建筑对谈录

李晋　编著

中国建筑工业出版社

李晋：

工学博士，华南理工大学建筑学院教授，博士研究生导师，

亚热带建筑科学国家重点实验室研究员，

2016 年中国建筑学会·青年建筑师奖获得者，

荷兰代尔夫特理工大学国家公派访问学者，

中国绿色建筑与节能委员会委员，

国家一级注册建筑师，

研究方向：大型公共建筑创作与理论研究、绿色建筑研究、健康建筑研究，

多次获得全国及省部级建筑创作奖项。

序　言

1 导言

在过去的半个多世纪里，荷兰现代建筑一直是全球建筑发展的引领者，尤其是近30年，荷兰建筑的爆发造就了"超级荷兰"现象，荷兰建筑师的作品得到国际社会的广泛认同与关注。代尔夫特理工大学建筑与建成环境学院是荷兰建筑教育的代表，笔者利用在代尔夫特理工大学做访问学者的机会，与多位荷兰具有代表性的建筑师进行了对谈，就其设计理念和作品进行了深入探讨。

通过对荷兰建筑的观察、与建筑师的对话，笔者对荷兰现代建筑的基本状况有了一些新的认识——其整体呈现出多元发散状态，面对所呈现的发散式表象，是对形式与空间的追寻，还是有更深层次的思考？是在现代主义建筑原则的基石上演绎，还是另有源头活水？——通过自己对访谈的整理和对建筑的切身体验，对西方主流建筑师创作的基石与思维方法的本底有一定的印证：既发现了彼此之间思维方式距离的原因，也看到了多位建筑师之间不同的表象之下的共同根源。由此，笔者深切体会到，一方面，荷兰建筑是深深根植于"荷兰性"基础上的，即荷兰建筑师对荷兰特性的深切理解与把握。虽然建筑师所运用的具体设计方法不尽相同，但都从不同的视角阐述着人与荷兰地域特征、环境特征的关系。另一方面，荷兰建筑近30年与几十年前创作遵循的共性理念（现代主义）相反，建筑师的创作理念多基于人的"体验性"——尊重个体对环境的感知与内在感受，由此，角度多样、个性迥异并且主题鲜明的作品得以不断发生出来。"荷兰性"与"体验性"

犹如荷兰建筑相互映衬的正反两面，正是荷兰建筑对人感知（内在感受）的尊重，荷兰的环境特性才能真正成为荷兰人内在体验中的映射，于是"荷兰性"才能自然而然地发生出来。"体验性"是共同的本底，而"荷兰性"是呈现的结果。

因此，好的设计应该能深刻体现人与环境之间感知与被感知的关系。建立人与环境之间的感知关系是当代西方主流建筑师的共同本底，形式与空间只是其推演的结果。"关注人的感受，尊重人的情感"，以此为出发点进行思考即为西方建筑师创作的基石。国内某些建筑实践方面，面临着在表面上跟风，但实质上却南辕北辙的窘态。这就是造成"奇奇怪怪"建筑并导致巨大浪费的根源。"尊重人的情感，关注人的感受"，并以此为出发点进行思考与演绎，才是真正的"以人为本"，也是建筑师应该思考的角度。"好"的标准不仅仅在于经典，也根植于建筑师自身真实的体验与认知。只有扎根实践，真实地面对自己的感受并进行拓展，也许很多新鲜的视角与逻辑可以由此发端。如何扭转热衷表象转而脚踏实地"认识环境与人的需求，服务于人的感知、服务于人与环境的关系"，我们依然有很长的路。

2 起点——一段体验与对话引发的思考

对这一问题的讨论，是从我刚到荷兰第一天的初次建筑体验开始——代尔夫特理工大学的大礼堂（TU Delft Aula）与图书馆（TU Delft Library）。这两个建筑位于大学的中部，彼此相邻。这两个建筑特色鲜明，反差强烈，而又非常协调。它们属于不同的年代，也基于不同的设计理念。显然，大礼堂的设计原理是在功能和技术之后的形式，呈现出粗野主义风格；而在图书馆的设计当中，可以找到人与环境之间更为直接的关系。

代尔夫特理工大学大礼堂（TU Delft Aula） 代尔夫特理工大学图书馆（TU Delft Library）

2.1 体验

大礼堂建成于 1966 年，由荷兰建筑师雅各布·贝兰德·巴克马（Jacob Berend Bakema）设计，粗野主义风格——东部的巨大主观众厅被粗壮的支柱高高架起，形成建筑的主体，粗犷有力的体量、裸露的混凝土烘托出建筑的纪念性格———一座雄壮的纪念碑式的建筑。主观众厅下部的架空层构成建筑的主入口。檐口强烈的水平挑檐将建筑的各个体量完整地统一起来。南北两侧结合楼梯形成挑檐，加之落地玻璃，形成开阔的景观平台。"*巴克马在20世纪60年代设计了大礼堂，它是完全不一样的，体现的是粗野主义的未来，它和混凝土的发明与建筑的结构有关。70年代时，我已经在这所大学里面学习了。我非常喜欢在这所大学里面学习，但是没有在这里，我也在建筑系的楼里学习过，它也是由巴克马设计的。老实说，我非常喜欢它，但是那里没有室外的空间，没有校园的感觉，建筑外也没有可以坐着的地方。*"*

图书馆建成于 30 年后（1996 年），由荷兰建筑师法兰馨·胡本（Francine Houben）设计。当时我在校园里先看到的是大礼堂，当绕过大礼堂，首先映入眼帘的是一片向上翻起的草坪。与大礼堂强烈有力的体量相比，图书馆表现出一种轻快而亲切的性格。当时自己被这种强烈的印象所推动，跑上草坪屋顶，代尔夫特理工大学整个校园的景色尽收眼底，那是一种无法言喻的心旷神怡！——在这里，似乎找到建筑对人的最为直接的关心，可以找到人与

* 本小节引文均引自代尔夫特理工大学图书馆设计者法兰馨·胡本的描述。

环境之间最直接的关系。"当我们开始设计图书馆的时候，我认为非常重要的一点是，巴克马的建筑就像是在月亮上的一艘船，因此你不能单纯地把另一栋建筑放置在这个建筑旁边。这会影响大礼堂的主导地位，我们选择将建筑设置在一个大草坪下，然后将草坪升高，这样就建造出了图书馆。这样的做法也形成了一个我一直渴望拥有的室外空间，你可以坐在这个室外晒太阳。假如我做了一个曲折的屋顶绿化，我想这只能让小狗上来，但是人走不上来。我认为不像巴克马的设计也会更好。这就是我为什么说'他们是父女'的原因了，'他们'现在是一家人了。"

2.2 思考

这两个彼此相邻的建筑均强烈触动了我，但似乎又存在于不同的层面。荷兰建筑的快速进展，绝不仅仅是停留在风格、形式或者建造技术层面的，建筑设计思考的"基点转移"是更深刻的因素。其内在因素是什么？背景是什么？将来的方向又如何？如果仅仅从建筑学范畴来观察，或者从谱系学的角度来讨论，都是就事论事，永远是雾里看花。因此，我们需要对自己的既有认识和认识方式进行反思，进行再认识。

我们一方面应该转换观察的视角，不应该仅仅从评论的视角和文本的视角，而更需要从建筑师的视角、亲历的视角。对我们来说，现代建筑是舶来品，我们的认识方法，不能停留在书本上对流派的梳理与文本式的理解、推断，而应该从建筑师的视角去亲身体验，说出自己的切身感受；应该用亲历的方法，去体验、辨别不同时期建筑所带来的直观感受，来印证不同时期作品的价值。只有通过亲历，才能做直接的验证。这一点在书本上是体会不到的，这是我们能提高对荷兰建筑认识的直接、有效、准确的途径。另一方面，应该将荷兰建筑放在更为宏大的坐标系背景（时间、范畴与地域）中来观察——第一，从时间的前点回看历史，历史的沉淀如地壳断层的纹理一样，层次清晰。将清晰的层次梳理清楚，不难推断出将来的方向。历史是由多个阶段连接而成的；每个阶段都有其积极因素与局限性，每个阶段都是对前一阶段的修正。第二，现代建筑与现代艺术的关系密不可分；那么，现代艺术处于什么状态？现代建筑在其中又处于什么位置？只有这样，才是用关联的眼光来看建筑。第三，从整个欧洲建筑的大背景来观察荷兰建筑，只有这样，才能将地域的眼光带入观察的视角。总之，只有从相互交织坐标系的大背景中来观察荷兰建筑，才有可能看清其来龙去脉，才能发现其真正值得学习的东西。

3 一百年的距离

历史是渐进式的，是由多个阶段连接而成的，每个阶段都是对前一阶段的修正。从感知的角度来思考现代艺术，到从感知的角度来思考现代建筑，经历了一百年的时间。

3.1 现代艺术的百年（1860年代—1960年代）：印象主义、后印象主义和抽象主义、表现主义

（1）印象主义

印象主义（Impressionism）在19世纪六七十年代以创新的姿态登上法国画坛，在光学理论和实践的启发下得以兴起，其绘画观念与方法具有一定的科学性，强调瞬息即逝的光色效果，追求感受光色。"他们想通过光谱式的色彩，来认识这个可见的世界，由此说明，物体之间的空间，也像物体本身一样，实际上是有色彩的间隔"。* 同时轻视文艺复兴以来的所强

《印象·雾（小港）》（莫奈，1872）

* 引自：西方现代艺术史，第35页。

9

调的三度空间中的三度实体。期间涌现出一大批绘画大师，如马奈、莫奈、毕沙罗、德加、雷诺阿等。从本质上讲印象主义依然属于西方客观再现的艺术传统。

（2）后印象主义

后印象主义（Post-Impressionism）是法国美术史上继印象主义之后，存在于19世纪80年代至20世纪初的美术现象，代表人物如塞尚（Paul Cezanne）、高更（Paul Gauguin）及凡·高（Vincent Van Gogh）等。他们不同于印象派追求外光和色彩，都认为绘画不能仅仅像印象主义那样去模仿客观世界，而应该更多地表现画家对客观事物的主观感受。他们主张重新重视文艺复兴或巴洛克大师们所强调的"形"的观念，并重视画家的主观体验，注意在作品中表现画家的主观感情和情绪，注意形式的表现力。正如《西方现代艺术史》所描述的"由艺术家的各种体验而形成的绘画，本身就是一种独立的真实"。"塞尚在他的作品中，所寻找的就是这种真实，即绘画的真实。他逐渐感受到他的源泉必须是自然、人和他生活在其中的那个世界的事物"。*这种"真实"构成了一个类似于但又不同于他所生活在其中的那个世界，这个世界中充满了画家对现实世界的体验与观点。"塞尚的绘画属于文艺复兴和巴洛克风景画的伟大传统，然而，又像眼睛看到的那样，它又被看成是个人知觉的极大积累。画家将这些分解为抽象的成

《埃斯泰克的海湾》（塞尚，1886）

《静物苹果篮子》（塞尚，1890-1894）

《圣·维克多山》（塞尚，1904-1906）

* 引自：西方现代艺术史. 第35页。

* 引自：西方现代艺术史，
第 35 页。

分，重新组织成新型的绘画的真实"。* 这种"积累"是画家个人体验的积累，从而构成了建构新图景的基本元素和灵感源泉。西方现代画家将塞尚称之为"现代艺术之父""现代绘画之父"。

（3）20 世纪上半叶的艺术

后印象主义引发了像雪崩一样古典主义价值观的崩塌，确立了崭新的价值观，被称为西方现代艺术的起源。后印象主义画家的探索过程与成果对 20 世纪西方现代美术流派产生很大的影响。后印象主义绘画偏离了西方客观再现的艺术传统，启迪了后来的两大现代主义艺术潮流，即强调结构秩序的抽象艺术（如立体主义、风格主义等）与强调主观情感的表现主义（如野兽主义、德国表现主义等）。这些探索在表现人类自身情感方面走得更远，影响至今。

（4）现代艺术的本质

回顾现代艺术发展的百年，可以看出艺术两方面的转变。一方面，表达主题的转变——从表达对象到表达自我。后印象主义画家主张表现自己的主观情感，色彩是否是物体的原来色彩已经不重要了，重要的是用这种物象来作为表达人类自身内心主观情感的媒介。因此，后印象主义画派注重如何在绘画中强调表现画家对客观世界的主观感受

** 引自：西方现代艺术史，
第 35 页。

"类似于但又不同于他们所生活在其中的那个世界"**，不再片面追求外光和色彩效果在画面上产生的真实感觉，而是具有更为主观化的感情因素和象征性的精神观念。这种观念的变革具有划时代的意义。正如《西方艺术史》所说"塞尚之所以能够建立一个新的绘画概念，并对二十世纪的艺术产生了开创性的影响，乃是靠了印象主义者们的色彩和瞬间幻想，以及古代大师们的训练和坚实的结构，更重要的是靠了他那观察自然的强烈而敏感的知觉"。***另

*** 引自：西方现代艺术
史，第 37 页。

一方面，表达方式的转变——从再现走向表现。塞尚主张绘画摆脱文学性和情节性，充分发挥绘画语言的表现力。他的作品注重理念和结构。他对体面的深入研究和高度重

视，孕育了立体主义的因素。后印象主义融入了个人的情感、思想，出现了史无前例的革命，强调探索个人情感对世界的体验，强调人的情感与世界的统一关系。从此艺术与再现分离，走上了表现的方向。

现代艺术自后印象主义开始，其本质是感知的艺术——强调表现"见所未见"，而非"所见"；是表现的艺术，而非再现的艺术。表现"见所未见"是区分现代艺术与古典艺术的分水岭——描绘体验到的，而非仅描绘所见的。这里说的第一个"见"，并非指眼睛看到的东西，而是感知到的东西；第二个"见"是眼睛看到的事物。这四个字说的是感知到了视觉之外的事物，看到你心中的图景与体验到的氛围。越是个性的越具有共性，越是个人的越是世界的。后印象主义的出现是挖掘在人类自身内在情感规律，这种趋势成为一种必然。因此，后印象对前是"继承"，对后是"拓展"，起到"继往开来、承上启下"的作用。从此，现代艺术走上了探求世界本质、表达自我和挖掘人类自身情感规律的道路，并一直延续至今。

3.2 现代建筑的百年（1920年代至今）：基点的多次转换，技术美学……

从塞尚开始，绘画已经不再仅仅表达视觉看到的东西，而是开始画感受的东西——这是现代艺术的真正起点。表面是画面的问题，而其实背后是价值观的问题，离开价值观来讨论艺术是没有意义的。作为现代艺术一部分的现代建筑的美学基点应该从哪里出发？视觉还是感知？现代艺术是感知的艺术，从感知与体验的角度来理解当代欧洲建筑的发展更能顺理成章。

（1）现代主义建筑的思想基础与方法原则：统一的价值观——技术理性与技术美学

现代建筑的前50年——现代主义建筑（20世纪20—70年代），是基于技术理性原则的美学表达。现代主义的思考方法，在完全尊重技术原则和技术特征的技术理性基础上的创作思考，包含了对技术固有特征的强化（粗野主义、高技派），对情感（集体观念或者意识形态）的追求是一种对象性的追求。

技术理性是合理地应用对实现目标所用的方法观念，基于环境和自身条件的限制，调整行为，排除偶然因素，以达到以最小的代价获取最大的收益。技术美学是以技术理性（合理性）为基础的美学设定。用当代的眼光来看，这一美学定义已经把美的范畴限定在合理性的范畴之内了。这也大大缩小了体验与感知的边界。

技术美学是以技术理性为基础的。在这一时期的建筑中，从粗野主义到高技派，首先是不同建造技术体系的转换（从混凝土到钢结构），而后是美学价值取向的转换（从粗野主义到高技派）。其中我们首先看到的是技术（手段）；技术（手段）是这个时代美学的显性载体。在人的感受和建筑之间始终有着一层隔离——技术；或者反过来说体验是隐藏在技术背后的东西，使人不能直接触及建筑感动人的一面。建筑始终在力图真实地体现技术的基本原则，无论自然而然还是刻意为之。这一美学框架的禁锢，一直延续到20世纪90年代。

（2）现代建筑近30年的思想基础与方法原则：平行世界中的感知与体验

现代建筑近30年（20世纪90年代至今），基于人的感知（情感）的理性（技术）异化。这一时期，技术原则

乌特勒支大学学习中心

的规则被打破，表现为以情感（个体的感受与体验）的追求为立足点，以技术的发展与异化为支撑点。

这个时代的建筑美学特征是：直接与人的感受与体验产生对话；而支撑建筑的技术体系退居幕后，技术不再是第一位的事情。这个时代，从建筑看到的不再是技术，而是与人的感知的直接回应。因此，技术在这个时代的建筑特征中成为美学的隐形载体。在人的感受和建筑之间，没有隔离了；具体表现为对技术体系的隐含与异化。当建筑的技术性后退的时候，建筑中的人性才真正呈现出来。从1990年代以来，以雷姆·库哈斯（Rem Koolhaas）、妹岛和世为代表的荷兰、日本建筑师的作品真正体现这一点。建筑不再是技术的直接呈现，而是与人感知的直接呼应，与人身体的相互交融。在这个意义上，建筑的造型与空间成为构思的结果，而非本质与目标；技术成为退居幕后的支撑。

判断技术异化合理与否的尺度与边界，有两个标准：一方面，其判断基准点是否有利于对人类情感的探索与表达；另一方面，是否完全背离技术原则走到建筑的反面。不同建筑师手中的技术异化程度不同，比如彼得·卒姆托（Peter Zumthor）和雅克·赫尔佐格（Jacques Herzoge）的比较，代表了两个极端，两者对于技术异化的态度截然不同。卒姆托主要是坚持在构造技术层面的异化，在构造层面表达技术逻辑的改造与转化，使之呈现一种与人的感知之间相契合的关系；赫尔佐格将这种异化扩大到结构层面，为了创造某种特殊效果，是结构逻辑、构造逻辑与材料逻辑共同发生异化，在建筑尺度的把握上更接近城市尺度，其中就有将图像扩大到城市尺度的做法。为达到某种令人炫目的效果，其对于结构、构造和材料运用方法的探索有时走到了技术原则的反面；妹岛和世的探索则呈现更为系统的特征，是结构、构造与材料的成体系演化，更像中世纪教堂的演变过程。

随着现代主义整体价值观的转变，这一时期建筑创作的基点是个体的感知，建筑的创作方向建立在建筑师个体的价值判断基础上。因此，整体上呈现出横向关联性较弱的发散状态。不同的建筑创作主体（建筑师）之间的联系较弱，虽然会运用相近的方法，但由于不同创作主体背景（本底）的差异性，观察世界角度的差异性，导致在创作的原点（出发点）上区别明显，呈现出个体对世界的认知与看法。因此，呈现出彼此"平行的世界"。

（3）后现代主义建筑的思想基础与方法原则：目标与手法的错位——情感的回归与脱离本体的表达

从 20 世纪 70 年代至 90 年代之间的 20 年，是价值观转变探索的 20 年。两者之间经历了"脱离技术理性原则的情感表达"和"完全依赖理论的情感表达"两个阶段。而这两个方向因为脱离了建筑的基本属性而不可持久，构成了前后两个阶段的主题回归与方法探索的"过渡期"和"探索期"。

3.3 从"现代艺术的百年"到"现代建筑的百年"——两个相互错位的一百年

美国建筑师菲利普·约翰逊（Philip Johnson）有一句名言："建筑是艺术这趟列车上的最后一节"。意思是说艺术发展的成果反映在建筑领域，一般都是相对滞后的。

1890 年代，后印象主义的表现创作者主观情感的观念被艺术界普遍接受。其标志性事件是 1900 年，莫里斯·德尼斯画了一幅《向塞尚致敬》，如今悬挂在巴黎现代艺术博物馆。画面上，许多人，包括博纳尔、雷顿、鲁塞尔、塞吕西尔和德尼斯本人，围绕在一个被他们视作老师的人周围。

1990 年代随着后现代主义、解构主义风潮的过去，以雷姆·库哈斯、彼得·卒姆托、妹岛和世等为代表的一批在这个时代崛起的建筑师的作品被普遍认可。其标志性事件是库哈斯于 2000 年获得普利茨克奖。他们的设计理念与方法不尽相同，但他们的作品具有共性——尊重人的体验与感知。这标志着新一代欧洲建筑师、日本建筑师设计理念上的整体转变：建筑师从对环境的认知出发，尊重人的体验与感受，将建筑中人的感受作为一切思考发端的起点，这成为普遍的共识。这标志着后现代与解构主义之后，建筑师的整体转向——从相信依赖理论，到从人的感受与体验出发。而在这种整体转变之前，渐进式的改变一直存在，从晚期的柯布西耶，到阿尔瓦·阿尔托（Alvar Aalto）、凡·艾克（Aldo van Eyck）、路易斯·康（Louis Isadore Kahn），等等。2000年作为分水岭，新一代建筑师作为一个整体，开始思考人在建筑中的切身体验问题，而不是用思想、原则或者文本等方法作为媒介。

这里我们可以有趣地发现，菲利普·约翰逊所说的"艺术的列车"确实很长，从后印象主义为代表的现代艺术转向审视挖掘人的自身情感（1900年），再到当代建筑朝着相同的方向转向（2000年），前后经历了整整一百年的时间！自此以后的欧洲与日本建筑摆脱现代主义理论的约束，以探求人的情感体验为基石，以建筑师的自我为核心，呈现出没有边际的发散状态——建筑师是对环境所形成的氛围的解读者、体验者与建构者——他们既是体验者，也是解读者。

3.4 荷兰现代建筑的百年——引领建筑思维转变的一百年

20世纪20年代，自现代主义形成开始，现代主义就是有针对性的，具体是针对工业化大生产背景下的复古主义问题时提出来的。欧洲有很强的人文主义传统，现代主义在反对复古主的形式教条的同时，将人文主义的传统也一起抛弃了。这是现代主义的所存在的巨大历史意义的同时，所具有的历史局限性。当现代主义的操作法则（国际式）被当作原则不断复制的时候，其局限性很快展现出来。当任何东西只保留形式而没有内容的时候，就没意思了——不能再引起人们的感受与共鸣，成了程式化的东西。

50年代，作为荷兰建筑师在国际上的代表，凡·艾克和巴克马是指出荷兰现代建筑新方向的人物。这时的凡·艾克同柯布西耶已经没有关系了，他从文艺复兴和非洲的原始艺术中来寻找重新出发的坐标系和原点，他引述文艺复兴大师伯拉孟特的名言"城市是个大建筑，建筑是个小城市"，来强调建筑是生活的反映，生活丰富的特质应反映在建筑与城市之中。凡·艾克的"结构主义（Structuralism）"理论的提出正是基于此，其影响从1950年代一直持续到1990年代，影响了包括赫曼·赫茨博格（Herman Hertzberger）在内的一大批荷兰青年建筑师。作为Team10的重要成员，凡·艾克与其他成员一起，将人类学、社会学，包括战后的通俗艺术纳入建筑学里面了，这不仅拓展了现代建筑的边界，而且改变了现代建筑思考的基点，形成了更宽广的视野。这一点在同时期的教育领域也有所反映，当时在代尔夫特理工大学的建筑历史教育中，会提到晚期印象主义的理念与思想，会提到这种基于体验的思考问题的方法。凡·艾克的阿姆斯特丹单亲家庭公寓得到了非常高的赞誉，其程度甚至超过了孤儿院，基于行为与体验引出入口空间，让亲历者感受到了建筑对人感受的尊重。位于海牙的小天主教堂，从立面构成到内部空间，更加强调光线与人，空间与人的关系，而非视觉法则。80年代凡·艾克的作品更不在乎细节与形式，这时的作品是与现代主义构图法则最远，而与生活最近。正是由于凡·艾克与追随者共同的反对与呼吁，后现代在荷兰的影响非常微弱。与凡·艾克相比，巴克马更像是传统意义上，现代主义发展脉络上的建筑师，提出"邻里单位"、"视觉组群"等理念，在"二战"后的鹿特丹城市社区重建中得以成功应用。他更大的成功体现在单体建筑上，代尔夫特理工大学的大礼堂和泰尔讷曾（Terneuzen）市政厅，均是其粗野主义的代表作；同时期也有青年建筑师对新现代主义、新理性主义的探索。

作为下一代的荷兰建筑师的代表，库哈斯经常会反思"现在通行的做法对不对？"很显然，相较于前辈们，他采用了更为激进的态度与方法。基于现代世界的可能性的整体价值观判断，库哈斯认为城市比建筑更重要，更关注城市，这是他与凡·艾克很大的区别；在建筑形体空间上，库哈斯受到20世纪20年代立体主义、构成主义的影响很大，更加强调设计中对人体验与感受的触及，其程度更大于他的前辈。他普遍质疑当下流行的套路化作法，这与凡·艾克对现代主义套路的反对如出一辙。他思考问题的方式和他的作品在深度与广度上深深影响了其后的整整一代青年建筑师。

阿姆斯特丹单亲家庭住宅

当代荷兰建筑师灿若群星，卡斯·卡恩（Kees Kaan）的极少与多元，MVRDV的直截了当……不可尽数，总体呈现出一种发散的发展状态。有的根植于现代建筑的核心价值观不断深耕；有的则着眼于探索建筑新的可能拓展其边界。他们之间很近，均因其是基于荷兰建筑百年变革所形成的整体思维背景之上；同时他们之间又很远，因其均有相对独立的价值观，是基于建筑师自己对环境的认识与角度，寻求着相对独立的答案。甚至身处同一城市，而完全不了解对方的想法。

阿姆斯特丹孤儿院

纵观荷兰现代建筑的百年，无愧于作为现代建筑的引领者。一方面，其实践与理论拓展了现代主义建筑的内涵，两者之间不是一个否定另一个，而是后者拓展了前者的关系。另一方面，当代荷兰建筑与其出发点的距离已经很远了，经历了多次思维基点的转换之后，展现出一种与技术同步的

海牙小天主教堂（Pastoor van Ars-Kerk）

无限可能。

4 新的立场

我们对"创新"有了更深的理解。创新不是表面形式的变异，而是对生活内容的发现、对于人内在感知和情感体验的探求。设计的创新不是主观随意，不是漫无目的的乱想，而是在对人情感体验的深度挖掘基础上的深度建构（表达）体现在建筑上，是以对感知的探求和对内的自省为路径的。建构是以人的情感体验感知为基础的；形式与技术的组织也是为此服务的。

由此可见，从现代主义建筑到当代建筑的创作基点转换了，由技术转换为感知与体验；现代主义建构的逻辑秩序是以技术理性为基础的，而当代建筑建构的逻辑秩序与感知、体验秩序相一致。这既是创新的基本立场，也是观察、学习当代欧洲建筑，尤其是荷兰建筑的基本立场。

为了让读者更好地理解对谈内容，作者在对谈内容之前加入了导言部分。

李晋

二〇二〇年秋日 于华园

目　录

导言　两种看待建筑的立场

一、感知与建筑本体

本体（noumenon; thing-in-itself），康德哲学中的主要概念，指与现象对立的不可认识的"自在之物"。这里本体是指构成事物的固有要素与特性。建筑本体是指构成建筑的形式与技术等要素的总和，是建筑脱离文学性和情节性，由自身构成要素所呈现的本来面貌。只有当建筑本体与所承载的内容、意义分离的时候，建筑本体与感知之间才能建立起某种紧密的联系，才能在体验的方向上进行演化。

二、两种看待建筑的立场

建筑是人与环境的关系，是人对环境的认知关系。这种关系是怎样的呢？用什么立场来看待这种关系呢？可以划分为两种类型，体验性立场与对象性立场。

（一）"体验性"立场，建筑以本体的方式存在，以唤起感知与情感的方式存在。建筑在人与环境之间建立感知与被感知的关系。强调感知、建筑本体与环境的三者统一。身在其间，会感受到强烈的自我存在感。基于"体验性"立场的设计方法，就是建构本体要素与感知之间的密切联系。基于体验设计思考包含了自身认知的独特性，是一个自然生长的过程，一个因人而异的过程，一个强调自我意识和自我体验的过程。表达方式上是属于表现的范畴。在这过程中，理论与技术体系成为工具，而非价值观。

（二）"对象性"立场，建筑以工具的方式存在，强调将建筑作为叙事的工具，强调建筑的文学性和情节性、强调表达建筑本体之外的内容与意义。建筑作为一种对象性存在。基于"对象性"立场的设计生成过程，是以理论、风格作为设计思考的原点，以工具来认识环境与生成建筑。越接近范本，就越接近标准答案。表达方式上是属于再现的范畴。

前者强调感知与建筑本体的密切关系，强调体验；后者强调建筑文学性与情节性，将建筑作为一种叙事的工具。两者在实践中往往存在互补性。两者的最大区别在于认知的立场，前者是站在人的立场上，后者是站在物的立场上。前者是建筑与人的一体化存在，后者是对象性存在。从体验性立场来认识当代的建筑发展，是基于建筑本体与体验的密切关联，是发自本心，而非原则；讨论问题的方式发自感受，而非条框。在当代欧洲实践中，大部分建筑师更倾向于前者，这也说明了欧洲建筑整体呈现发散状态的根本原因。

三、感知的多向性与自我认知

人的感知既有趋同的部分，也有因人而异的部分。对于建筑师而言，在实践中认识了解自我，表达自身的真实体验，是建筑向着开放体系发展的重要前提。而勤于实践是建筑师自我认知的重要途径。对于现代主义的认识应该站在历史的高度，对其原则与结论，既要了解其历史意义，也应认清其局限性。理论既是桥梁，也是屏障。

四、感知的多向性与创作的发散性

当代建筑的发散性是立足于体验性认知基础之上呈现的自然状态。一方面有共同的本底，另一方面又有建筑师自我认知与探索，整体趋同的风格消失了。很多建筑师的工作方法多是从建筑本体要素与体验的关系来讨论问题，这在日本建筑师中体现得尤为明显。

对人类情感广度与深度的探求，正引领着建筑的发展，并呈现出以感知为基点的发散状态。对于建筑的繁荣，并非只是标准化，而应该是多样化。多样化并非是形式的多样化，而是体现感知的多样化，而是动人氛围的多样化。

五、设计触及感知的过程

什么能触及感知？氛围。什么能制造氛围？建筑本体的一切要素。立足于对人的研究，来思考建筑问题；而非脱离人，单纯地思考建筑。

首先，环境中的感知与体验是设计思考的原点。能打动人的是建筑所形成的氛围。基于环境体验，建立何种氛围是设计中首先应该思考的。进而，用什么技术去营造这种氛围？思考建筑本体要素（如结构、材料、尺度、质感因素等）与氛围的关联性尤为重要。根据氛围的需要来组织对建筑要素的建构。法兰馨·胡本在谈到创作代尔夫特图书馆时，着重描述的是创造宜人的室外空间的理由——在多雨、阴冷的荷兰环境里，创造了倾斜向阳的草坡，创造了温暖的室外空间；瓦尔斯温泉，条状的石材机理、细长的光线从石缝间泄下、狭小的浴室中声音的回响、透过框镜所看到的远处的群山，这一切因素都在拨弄着观者的心弦。在这里，彼得·卒姆托对材质与光线的表现可谓淋漓尽致。随着氛围的建立，这一思维过程的最终呈现为形体与空间。形体与空间是用来制造"氛围"的手段，是结果，而非本质。在这一过程中运用的理论与方法，作为可以反思的工具存在，而非价值观，价值观存在于氛围层面。当代日本建筑师的思维方式也大多在这一范畴。

随着社会的改变和人们视野的拓展，感知的领域在快速扩大，制造的氛围也是空前丰富，最终呈现的形体与空间不断带给我们惊喜。因此，西方建筑的形式的演进与变化，其内在动力并非形式本身，是内在因素的不断变化。从感知与体验出发，创作出动人的氛围，是我们的最终目的。在创造氛围的过程中，呈现为形式与技术的进化过程。

沿着这一思维逻辑推导下去，会发现形体与空间并非本质，而是手段与结果。如果我们不理解这一点，而仅仅关注这些表象的时候，自然感到迷茫和无所适从。

六、感知与理论

感知与建筑本体之间相互作用的演进过程，方向是向内的，审视自我的，通过感知去寻找能够触动人的内在原因。从这个意义上讲，设计是触及感知的过程，建筑是触及体验的媒介。当建筑师们遇到感知与既有理论的矛盾时，应当如何选择？是跟从前者还是后者，取决于个人的价值观，而感知是带领我们继续向前走的根本动力之一。

七、感知与技术——技术边界的异化

技术是建筑发展另一条腿。现代主义的建构逻辑是以技术理性为基础，而当代建筑的建构逻辑与体验产生的过程相一致。前后两个时期，创作的基点转换了，由技术理性转换为感知与体验。

现代主义时期，技术原则的边界就是建筑的边界。现代主义的建构逻辑表现为，真实表现结构与技术的固有特征。将结构的力学体征、材料特征进行极端地表现，使其成为美学特征，如粗野主义的代尔夫特理工大学大礼堂和高技派的蓬皮杜中心等。

当代建筑的建构逻辑表现为，将技术理性与感知结合起来进行综合思考。在当代建筑中，随着创作的基点转换，对技术原则的遵守出现了异化。异化的根源在于不是为了理性而理性，而是为了情感而理性，对待理性的基点变了。不再将技术合理性做唯一的基础条件。牺牲技术的局部合理性以达到某种效果成为常态。这不是颠覆技术理性的过程，而是突破其局限性、拓展其外延的过程。当前，总体上技术发展的进程是落后于对感知的探求，对技术理性原则的突破成为必然。当前不同的建筑师对待技术的态度产生了分化。有些建筑师强调在尊重技术原则的基础上探索感知与技术体系的契合关系，如彼得·卒姆托、西扎等人；有些建筑师则是从较大程度上颠覆技术原则，对技术材料更为夸张的使用来制造感知错觉的过程，如赫尔佐格、妹岛等人；有些建筑师强调从更大的社会背景来思考，创造者更大范围生活布景的过程，如 MVRDV 等。

八、结语

设计创新不是主观随意，不是盲目跟风，而是基于生活视角的发现，是对感知和体验的探求，是基于情感的深度建构。因此，对情感体验的表达是以对内自省、自我发现为路径的，形式生成与技术组织也是以此为路径的。对于建筑创作的繁荣，我们并非只需求标准化，而是需求多样性。并非是样式的多样性，而是能够体现感知的多样性，而是动人氛围的多样性。感知的差异性，正是提供多样化的根本动力。

引领世界的荷兰现代建筑就如一座巨大的宝藏，有很多东西等待我们去发现。不同的观察角度，看到的东西不一样。基于在荷兰的学习以及回国后的印证，笔者梳理出一个"体验性"视角，希望这个视角能为读者提供一个再认识荷兰建筑的新视野。

对谈卡斯·卡恩（Kees Kaan）

访谈时间：2016 年 9 月 11 日

卡斯·卡恩是代尔夫特理工大学
建筑与建成环境学院教授、博士研究
生导师、建筑系系主任、荷兰卡恩建
筑设计事务所的创始人。他在荷兰以
及其他国家都成功设计并建成了许多
建筑作品，同时荣获了不少国际奖
项。其中荷兰最高法院荣获 2017 年度
Fassa Bortolo 国际可持续建筑银奖；伊
拉斯谟大学医学教育中心荣获 2015 年
密斯·凡·德·罗提名奖。

本章就卡恩教授设计的四个建筑
与其进行了讨论，分别是荷兰最高法
院、阿姆斯特丹法院、伊拉斯谟大学
医学教育中心和荷兰鉴证中心。

卡斯·卡恩（下文简称"卡恩"）：在过去的一段时间，你都去过哪里了？

李晋（下文简称"李"）：我去过瑞士，参观了几位重要建筑师的作品，一个是由雅克·赫尔佐格（Jacques Herzog）和皮埃尔·德梅隆（Pierre De Meuron）设计，建筑外立面很模糊，也非常有趣。另一个是彼得·卒姆托（Peter Zumthor）在库尔的作品，我认为他的作品给人很强烈的感觉，作品充满了感情，真的很棒！我在巴黎看过克里斯蒂安·德·鲍赞巴克（Christian de Portzamparc）的作品和让·努韦尔（Jean Nouvel）的作品，他们的作品完全不同。

卡恩：是的。

李：当我站在鲍赞巴克设计的巴黎欧风路 209 户住宅前，那是一组白色的建筑，有很大的体量，但它却被切割成若干个小的片段，因此当我站在它面前，不会感觉到压力，而是觉得很亲切。我认为它是基于城市的理念，基于人与建筑体量，以及人与建筑之间距离的关系，这种关系无法从照片中感知。整个体量被切分成很多小的部分，站在这个建筑前面的时候，人会觉得很亲切，不会感觉到压力。努韦尔的建筑，他的空间很模糊，很轻盈，很不稳定，很有趣。两个建筑是完全不同的，方向不同，方法不同，但我认为它们也有共通之处——可能是对人的感知与情感的尊重。但是怎样才能营造这种感觉呢？每个建筑师都有不同的方式。我认为这个问题很有趣。

KAAN: In recent time, where have you been?

LI: I have been to Switzerland and seen works of two important architects. One was by Jacques Herzog & Pierre De Meuron, the facade of their building is ambiguous and very interesting. Another one was in Chur mountains, by Peter Zumthor. I think his work leaves a very strong perception, and his work is full of emotions. It is really good. And I also went to Paris where I saw Christian De Portzamparc's and Jean Nouvel's work, they are totally different.

KAAN: Yes.

LI: When I stand in front of the dormitory of Portzamparc, a white house, it seems as if it should be a big volume, but it has been cut into several little parts. When you stand in front of it, you don't feel strong pressure, you will feel very comfortable. This relationship between the volume of a building and the distance from a person to the building - I think it comes from the idea of the city, this kind of impression cannot be obtained from pictures. The whole volume is divided into many small parts. When standing in front of the building, people will feel very kind and will not feel pressure. Jean Nouvel's space is very ambiguous, narrow, light, unstable, it is very interesting. I think they are totally different, with different directions, different methods, but I think they have things in common. I think it's maybe the respect for a person's perception and feelings. But how to evoke these feelings? Every person has a different way. I think the question is interesting.

卡恩：是的。

李：我还去了西班牙，那里建筑师的设计方法有很大的不同。此外，我也参观了你五个作品，都很有趣。其中一个在海牙，是刚刚建成的荷兰海牙最高法院。我想谈谈这个建筑带给我的第一感受。

卡恩：好的。

李：我在两个月前参观了这栋建筑，第一感觉是建筑深深触动了我，这个建筑很安静也很漂亮，而且新的建筑和旧的环境有很紧密的联系，它对潜在的环境有一种更清晰的表达。此外，当我站在建筑对面的街道上，面对这个建筑从远处看它。在夜晚这个建筑变得非常优雅，几乎成为一种背景。你可以看到建筑前树林的轮廓变得非常清晰、美丽。我可以想象得到当站在门厅里向外看时，你可以看到美丽的风景。当你在建筑旁边看建筑外立面上的玻璃时，可以看到树的轮廓倒映在玻璃上。建筑的每一个部分均能触动观察者的神经，因此我对在设计过程中的思考方式非常感兴趣。你对潜在环境的感知是建筑设计中重要的部分。您的感知、对旧环境的感受是不是非常重要的出发点？

KAAN: Yes.

LI: I went to Spain, all the buildings were different. I have visited five of your works in the Netherlands too, they are all interesting. One is in the Hague, recently finished - the Supreme Court of the Netherlands. I'd like to talk my first impression about this building.

KAAN: OK.

LI: I visited the building two months ago, my first impression was that it is very quiet and beautiful, there's a close relationship between the new building and the existing environment, in fact, as if it makes the context clearer than ever before. I went to stand on the opposite side of the street, in front of the building and looked at it from distance. During the night, it became a gentler, almost blending into the background. You could see the tree outlines in front of the building turning clear and beautiful. I can imagine that when you stand in the entrance hall and look outside, you can see a beautiful landscape. And when you are next to the building looking towards the facade, you can see the tree outlines reflected in the glass. Every part of the building can touch the nerves of the observer, so I'm very interested in the way you thought during the design. Your perception of the existing environment is an important part. Is your feeling and perception of the old environment a very important basic idea?

卡恩：你的观察非常细致，也很敏锐。对于我们来说有一点是非常重要的，即建筑所在街区的信息——例如树的位置、街道的位置——界定了这个项目。另外，这个地段在城市中的重要性、城市的历史文脉，都影响着我们设计这个建筑的方法和过程。这类项目在荷兰是非常重要的。城市扮演了一个非常重要的角色，因为它明确规定了在具体的地方能够建造什么建筑，所有的建筑都被给予了一个可建范围，建筑物的规模和建筑物的高度等都是被城市规定的。这在这里是通行惯例。对于你能建造什么，或者你被鼓励去做什么，都有非常明确的规定。但是，在可建范围里，设计一个建筑还是有很多的可能性，比如如何组织建筑的空间，以及如何将它们联系起来，如何体现你的喜好，如何把这些跟公共空间相联系，把入口设置在哪里，把雕像放到哪里，赋予建筑什么材质。对我们来说，项目所在的街道 Korte Voorhout 非常重要。Korte Voorhout 街道很短，一侧连着开放公园，另一侧则是政府的特殊区域和大使馆，这个区域的特质是空间被树林所覆盖，这是设计关注的一个重点。另一个重点是，设计要考虑最高法院本身的特点，最高法院的组织具有二元论的特征。一方面，它是一个非常重要的公共机构，最高法院的裁决影响整个国家的司法体系。法院裁决对社会意义重大，是极具公共性的典型职能，你可以说它根植于整个社会。另

KAAN: I think your observations are really on point. This was very important for us: the situation and position of the building in the literal sense - physical position of the street, trees, and space - has informed and defined the project a lot. The importance of that place in the city context and history has also played an important part in the way we organized the building. This kind of project is very important in the Netherlands. The city plays a very important role here because they are very strict about what one can do on a certain site, so all the architects are given the same envelope, the size of the building, the height, everything is fixed already by the city. That is something normal here, there are always very strict rules about what you can build or what you are stimulated to do. But inside that envelope, there are still lots of different possibilities to develop the project, regarding how you organize the program in the building, the spaces, and how you link them together, also in terms of materialization, expression, and how it all connects back to the public space - where you put the entrance, where you put the statue, the kind of quality you give everything through the material... For us, this street was very important - the Korte Voorhout. It's a very short street and creates a connection between the big open park on one side and the 'chic' area of the government and all the embassies on the other side, the characteristic of that space is that it has a green 'roof', the tree line, this is an important point of design. We took another important point from the program of the Supreme Court. Its organization has a very dual character. On one hand, it is

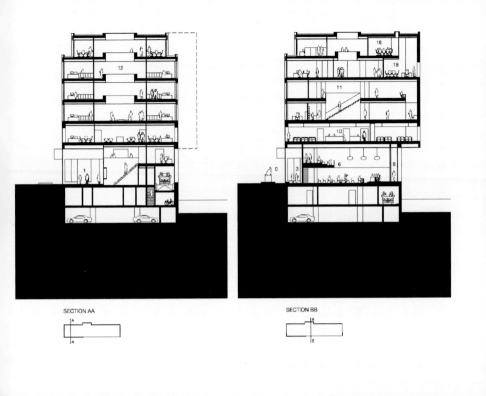

SECTION AA

SECTION BB

一方面，法院所做的工作是非常学术和抽象的，法院的工作人员研究很多事情，需要一些私密的空间，使他们能集中注意力。这是两个不同的方面。我们在做这个建筑的时候，给建筑设计了一个石质基座，仿佛建筑扎根在城市里，同时在基座上设计了玻璃的外立面，玻璃看起来非常抽象，可以创造一定的透明性。石材和玻璃的运用带给人亲切感，同时也产生了一定的距离感。就像法官的法官袍和假发——它们就像一件套装——代表一种机制，深深地去人格化。亲切感和距离感相结合，组成了整个建筑的性格。石头的部分是在地面上的，公共空间、建筑的前厅和走廊，一切都是从物质中切割出来的，营造了开放的空间。

李：是的，从大厅到外部我感受到非常强烈的引导。

卡恩：建筑与树高度相近，假设你站在建筑内向外看，犹如置身树下，这种感觉是对原有的公共空间有皈依感。

李：我们可以清晰地看到雕像的轮廓。

卡恩：是的。（荷兰司法史上原最高法院的重要人物的）雕像设置在外立面。你会感觉到整个司法过程与公共空间紧密连接。这既是一种象征意义的，也是字面意义上的。这与现代建筑手法无关，我们没有使用纪念性的或形式上的方式，而是运用了非常克制和简单的做法。

a very important public institution - whatever the Supreme Court rules can impact the entire bureaucratic system of the country; The rulings are very important for society, and they have a very public representative function; You could say the Supreme Court is rooted in society. On the other hand, the work they have done is very scholastic, very abstract; They study a lot and do that in seclusion, focused and concentrated. These are two very different things. So, we gave the building a stone base, as if it is rooted in the city, and put the glass on top of it. The glass top is very abstract, it can create a certain degree of transparency. It is inviting but also creates distance, kind of like the toga and the wig of the judge - it's like a suit - it represents the institution, depersonalizes it. Those two elements become the building. The stone part is on the ground; the public space, the foyer, corridors, everything is carved out of that material, so that creates the open space.

LI: Yes, it is a very strong connection from the entrance hall to the outside.

KAAN: And then the building is the same height as the trees. So, if you stand inside the building and look towards the outside, you're likely to be under the trees, this feeling is a conversion to the original public space.

LI: We can see the outline of the statues clearly.

KAAN: Yes. And then with the statues outside. You feel like the whole juridical process is connected to the public space. It is a bit symbolic, but also very literal. It is done with modern architecture; not with monumental, formal gestures, but with very restrained, simple means.

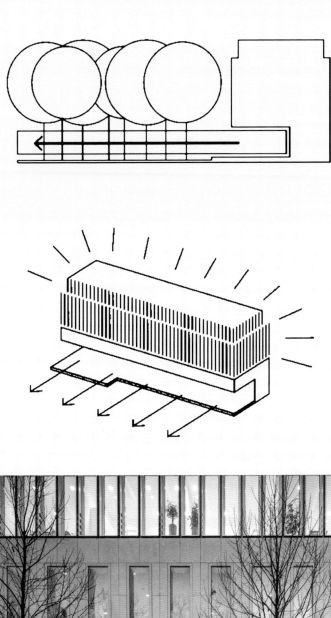

李：是的，建筑的氛围和旧环境非常契合，我很喜欢。

卡恩：我们从青铜雕塑开始着手，加上绿色的树，我们也选择了一种微微偏绿的石材，同时玻璃也带有一点绿色，所有的事物都产生了关联。

李：我认为你的分析非常准确，很细致，对所有元素的分析都非常有深度，能触动人的感知，并阐释你的想法。你的设计非常温和，甚至成为旧环境美丽的背景，同时建筑还唤起了潜在的环境氛围。您有一句话概括了您的想法："建筑的透明性意味着公众可以接近，也意味着判断的正确性和清晰性"。可以解释一下吗？

卡恩：它体现出一种透明性，人们可以看到建筑内部发生了什么，建筑内是没有秘密的。在荷兰社会，得到信任是很重要的，这是一个有趣的社会。

李：它非常开放和安全。

卡恩：当你走在街上，你可以看到城市的夜晚，也可以看到建筑内部，人们不会拉上窗帘。有一种观点是如果你没有要隐藏的事情，你就不需要把窗帘拉上。我认为高级法院也要这样，没有什么是需要隐藏的，它是完全开放和透明的，就如同他们裁决的方式，所有事情都需要讨论，他们所决定的事情都是非常清晰的。这里不需要窗帘。

LI: Yes. The atmosphere is very suitable to the old environment. I like it very much!

KAAN: It all started with the green statues, the green trees. We also choose a greenish-grey stone, and the glass also has some green to it, everything came together.

LI: I think your analysis is very accurate and careful. A very deep analysis of every element helps shape people's perception and expresses your ideas. Your design is very gentle, it blends in beautifully and evokes a certain atmosphere in the context. On your website you describe your design saying that:" The transparency of the building signifies both accessibility to the public as well as the soundness and clarity of judgment."

KAAN: Yes, it's literal transparency where people can simply see what is happening inside the building and there are no secrets. This is important to earn trust in the Dutch society, it is a funny society.

LI: It is very open and safe.

KAAN: Yes, for example in cities, if you walk along the street at night, you can look inside the houses everywhere, people don't close the curtains. There is this idea that if you have nothing to hide, you don't close your curtains. And I think the Supreme Court is doing that as well. There is nothing to hide, they are operating in complete openness and transparency. In their judgment and in the way they behave, everything is discussed and what they decide is an honest, clear decision. No curtains.

李：您的想法非常有趣。您的设计背后是否有一个强烈的意义？是否想去表达一种意义？是否尝试富于逻辑性地表达这些意义？有些建筑师极力去避免触碰意义，他们仅仅想去构建建筑物和人的感知之间的联系，但有的建筑师则恰好相反。你是如何思考这一点的？

卡恩：在这个设计里，我认为意义是非常重要的，最高法院在国家里是非常重要的一个设施，这是我们的民主制度的基础。

李：我明白你的意思。

卡恩：城市是我们个人的一种价值产物，反映着我们共同的想法、共同的喜好。这就是我如何观察我们社会的。基于我们的价值观，我们不断地建造，在这过程中塑造了我们周遭的环境。

李：您的意思是我们不能抽象地谈论某些事情，而应该根据不同的环境来思考。因为建筑是非常重要的，所以您赋予它含义。

卡恩：是的。对于普通的办公建筑，我们会以不同的方式来设计，但也是花了很多心思来设计。无论我们设计的是一个学校，还是公益住房或最高法院，我们都会花精力细心设计，用同样的关注度，但是着重点会在不同的地方。

LI: I think this is very interesting. Is there a strong meaning behind your design? Do you want to give it some meaning? Do you want to express some ideology? Because some architects try to avoid giving meanings, they just want to build a connection between buildings and feelings of people. But other architects want to put some meaning in it. What do you think happens in your projects?

KAAN: In this project, we wanted to give meaning because the Supreme Court is a very important institution in the country. It's fundamental to our democratic system.

LI: I understand what you mean.

KAAN: And the city is a sort of a product of our shared values, the city is the reflection of what we collectively think, of our common interests. That's how I look at our society. We build again and again, so we shape the world around us according to our values.

LI: So, you're saying that this cannot be talked abstractly, but instead we should think according to different situations and circumstances. In this case, because of the importance of this building, it needs to have meaning.

KAAN: Yes. Any normal office building, we'd treat it differently, but also with care. Always with care. With different importance, but with the same care. So, if we do a school, or social housing, or a Supreme Court, we always pay attention and always care. We just give it a different importance.

李：是的，我明白你的意思。建筑的上部有小中庭，四周围绕着餐馆、办公室和会议室，我觉得你对材质的运用很特别，墙和地面的材质是天然石材，这和白色的顶棚形成了强烈的对比，顶棚是白色的，而墙面是石材。在现代建筑手法里，墙面和顶棚通常是同一种材质，但这里不一样了。这是否是一种演变，从强调一种盒子的体积感，转变为强调一种轻快的感觉。

卡恩：我并不这样认为。我们并没有好像众人看到的那样从现代建筑思想上得到很多。* 我的意思是，现代建筑是一件非常普通的事，一件很成熟的，已经被接受的事。现代建筑在我们的社会中，在我们的环境下，是一种语言，而不是一种意识形态。** 在建筑上部的办公区域，我们想去做一个核心的办公区域，它有着不同的用途。这些办公室一部分是理事会，一部分是诉讼庭。它这两个实体相互补充，共同组成了最高法院。在它们之间是图书馆、餐厅和其他空间。在 16 —17 世纪时海牙的大宫殿里，这个部分之前都是围合起来的，拥有巨大的楼梯，楼梯和地板上的空白和浅白色天然石材，石膏顶棚，可以是说

* 卡恩教授是在反复强调，虽然荷兰当代建筑的源头是现代主义，但经过了多次演进之后，当前荷兰建筑师的设计出发点与 20 世纪的现代主义建筑原则之间的距离已经很远了。现代建筑不是一个静态的事物，而是一个不断发展变化的过程，包括其中的基本原则。

** 现代建筑在当下荷兰是一种常识性的存在，成为社会所普遍接受的事实，而非一种观念反对另一种观念的手段或者一种口号性的东西。

LI: Yes, I understand what you mean. There are small atriums in the top half of the building, surrounded by the restaurant, offices, and meeting rooms. I also think your use of material is special, the wall and ground are of natural stone, and it's in contrast with the white ceiling. In our modern architectural principles, walls and ceilings are made from the same material. Is this an evolution from emphasizing the volume of a box to emphasizing a sense of lightness?

KAAN: I don't think we are as much into the idea of modern architecture as people assume. I think for us, modern architecture is something very normal, something already accepted. It's more of a language than an ideology in our society and our environment. So, in the office part of the building upstairs, we wanted to do a very long sequence of offices. They have different uses - some parts are for the council, some for the prosecution. Those are two entities that make up the Supreme Court, they are two complementary departments. In the middle, there's a library, a restaurant, the president's room, and all that. They used to be housed in these big city palaces of the 16th and 17th century in the Hague, where they had big monumental staircases with voids, light, white natural stone on the floor, staircases

宏伟的宫殿。你可以说我们试图把皇宫里曾经有的堂皇和纪念性，隐喻到人们相聚交流、喝咖啡的空间里，与一众办公空间形成一种对照。通过这样的处理，我们希望建筑能给工作人员一种身份和尊严。我们没有特意从形式或现代建筑语言等角度来思考，我们已经完全从那里解放出来了。

李：我所感兴趣的是一般来说有两个不同的表面，以前材质应该是一样的，而现在，在许多当代的建筑中，建筑不同的侧面使用不同的材质。我感觉非常轻盈感的是，那不是一个盒子，而是一个表面和另一个表面的关系。我感觉不到重量了。感觉是……

卡恩：无拘无束的。

李：是的，无拘无束，我在建筑里感觉非常放松，而不是被强烈的力量所影响，仅仅是让我放松，我非常喜欢它。

卡恩：我能够理解。操作这类事情是有可能的，我们不是在一系列严格的规则下工作，我们可以自由地改变建筑表面的材质，在同一建筑中它可以从玻璃变成石头，等等。

and the walls and then - a plaster ceiling. You could say that we wanted to bring the same monumentality and grandeur those palaces had into a contemporary office and environment. Exactly in places where people meet, in the collective spaces in the building, not rooms where people work, but in the space where they can see each other and have coffee... They get a sense of this quality one could find in Dutch palaces ages before. By doing that, we give them a clear identity, an upgrade from the generic. Those very generic offices, they get an upgrade in the collective space and become more dignified, more important. And this has nothing to do with the idea of such formal notions about architecture or the modern architecture. Nothing what so ever. We are completely freed from that.

LI: What I find interesting is: generally, there are two different surfaces. According to old principles, the material should be the same on all sides. But now, in many contemporary designs and buildings, different sides have different materials. I find it very light, it is not a box, but an assembly of surfaces. It doesn't feel overpowering. It feels...

KAAN: Liberated.

LI: Yes. Liberated. I feel very relaxed in them. They don't overpower me but give me ease, and I really like it.

KAAN: I understand that. It's possible to play with these kinds of things. You know, we are not working within a set of rules, we have the freedom to change the surfaces in the material. It can go from plaster to stone, in the same building.

李：非常感谢你，我受益良多。我想聊聊的第二栋建筑是不久前中标的新阿姆斯特丹法院。它面积很大，比上一个建筑要复杂得多。通过它的剖面，就可以看出它非常复杂，从建筑一层到二层一直到顶层，每一层都有不同的层高、进深和空间水平面，同时在建筑外表面上也设计了不同密度的竖向构件。每个空间都有不同的层高、朝向和外观，还能看到不同的风景。你的目的是让人们看它的时候有不同看法吗？因为里面的空间是如此不同。一层外立面更宽，到了最高层又变得更加密集，你是根据人们所处的不同的环境创造了不同的感知空间吗？

卡恩：首先，总平面是非常重要的，我们希望广场能成为一个景观，我们把它拆分成四部分，然后将其中一部分挖下去，使它变成一个空间，一个广场。提供给城市一个公共空间对于建筑来说是非常重要的。因为人们会在这里进进出出，使这里变得很拥挤，我们需要去创造一种过渡空间。这个空间不可能放在周边城市环境中，因此它设置在法院建筑的用地范围内，并向公众开放。

李：提供给城市。

LI: Thank you very much, I've learned a lot. The second building I want to talk to you about is the New Amsterdam Courthouse, the competition which you just won. I think maybe because it's a bigger building, it is much more complex than the last one. From the section, I see it is a very complex space, from the first floor to the second floor to the top floor, each floor has different heights, depths and levels of space, and also different densities of the grid on the facade. Each space has a different height, orientation and appearance, as well as different views. Was your purpose to create different perceptions for the people looking at it, because the space inside is so varied? On the ground floor, the facade is wider, and it gets denser at the top. Did you create different perceptions according to the environment?

KAAN: First of all, the plan is very important. We hope the square to be landscape and divided it in four, then took out one part. This became a public square. It is very important for this building to give this space to the city. People go in and out of the building, it is very crowded and full. We needed to have this transition realm, which could not have been put in the surrounding public space. So, the square is a part of the court building, and it was given to the public.

LI: Given to the city.

卡恩：这是非常重要的。它是这个建筑重要的一部分，此外也是一种非常简单和谦逊的处理方式，这是一个很好的空间。当然，这个建筑内有一个法庭区，有两层供人们可以开会、讨论使用。五楼是供工作人员使用的公共层，人们可以在此交流会面，并可以在餐厅就餐。在这上面是工作空间。这个建筑的功能是法院，人们坐在里面等待，进进出出，和其他人相遇，和自己的律师见面，这就是在这个建筑中发生的事情，因此所有的方向你都要做成透明可见的。在底层法庭区，我们设计了一个大窗户，因为那是非常公共的空间。再上一层是人们工作会面和讨论的地方，虽然没有那么公共了，但也有一定的开放性。在高层办公区，你看到一些私人的办公室，那部分我们把外立面做得比较狭窄，对一部分空间进行了遮挡，但也有一部分是透明的。我们可以通过调整窗墙比来控制建筑的透明度。那个尺寸的窗户上有门，你可以轻易地打开窗户，移动它。这使窗户在某种意义上更加亲切了。在内部，这是可以理解的尺度，窗户是可以打开的，这很重要。

李：是的，上部的窗户可以推拉。

KAAN: That was a very important gesture, it gives certain importance to the building. And again, it was done in a very simple, very polite, quiet way, it's a nice space. And then, the building has courtrooms, it has two floors where people have meetings, conferences. The 5th floor is a very public floor, only for the workers to meet. People can meet each other here and have dinner in restaurant. On top of that are the workspaces. The foyers of the courtrooms are where the function of the building is displayed - what's going on in the courtrooms; people sitting, waiting to go in, coming out, meeting each other or their lawyer... This is something visible all around the building. In every direction, we make it clear and visible. That's why we have big windows on this layer - it is very public. On the next layer, it's where people work, meet, and have conferences, so it's much less public but still with an open character. If you go higher, you have private offices. There we made the facade narrower, it closes a bit more, but it's still transparent. And this allowed us to make windows that people can open, because of the proportion of the facade. We've reached the scale of a window that has the size of this door, so you can simply open it, move it. The windows downstairs are too big for people to handle. So, it also became more friendly in that sense. From the inside, suddenly, it's a scale we can understand, a window that can be opened, which is something important.

LI: Yes, the upper window of the building can be pushed and pulled.

卡恩：在心理学上这是很重要的，办公室应该有不同的尺寸。当你坐在比较小的房间里，桌子还有其他所有外界事物都让你感觉更具私密性。

李：这很重要。

卡恩：过渡空间的规模，从公共性到私密性的过渡，在外立面上就有所表达。

李：是的，我觉得你设计的外立面有不同的密度，它表达了不同的空间个性。

卡恩：而且外立面是承重的，或者说承重柱整合到立面里。它们支撑了建筑，在外立面上是完整的。承重柱不是在建筑的内部而是在外部。这意味着建筑内部看不到那些结构，没有突出来的结构，而是很平整，因此建筑内部是非常规整、干净。

李：我认为您是一个很有深度的学者。我认为你创造了不同特性的感知和空间，不像某些建筑师，特别是一些流行建筑师，他们总喜欢以流行的方式进行设计，我认为那是完全不一样的。

卡恩：我们不在乎那些流行的事情。

李：我认为建筑应该是对社会有一定的责任的，而不仅仅是表现个人的想法。

KAAN: It's important also psychologically, that in the office you have different scales, smaller rooms, with a desk... Everything becomes intimate and more private.

LI: Important.

KAAN: This transition of scale, the transition from public character to private character is expressed in the facade.

LI: Yes, I think your facade of different densities, it expresses a different character of space.

KAAN: It's also interesting that the facade columns are structural, they carry the building. They are integrated into that facade, and they are not inside the building, but outside. That means on the inside, you don't have any columns. Everything is flush, very clean.

LI: I think you are a profound scholar. I think you created a special kind of space, not like other people, especially some popular architects, who design fashionably. I think it is totally different.

KAAN: We don't follow any fashion.

LI: I think architecture should be responsible to the society, not just to express oneself.

卡恩：对社会同时也是对人类，对那些在建筑里工作的人，我们在方式上要以人为本，思考一些平常的事情，思考那些人们天天在想的事情。人们使用这个建筑，建筑是社会的一部分。因此我们设计的时候一直在思考建筑里公共和私密，以及和城市相关的主题。同时也关于尺度 * 和抽象，以及不同等级的抽象的和实体的，牢固的，干净的和清晰的。

李：我明白了，谢谢。第三个建筑是伊拉斯谟大学医学教育中心，在过去的两个月里我参观了这个建筑两次，我喜欢这个教育中心。我认为用普通的方式思考，抽象的空间是被地板、墙和顶棚所定义的。但是在这个教育中心里，空间是通过顶棚上的天窗来进行控制，天窗对整个建筑的氛围影响很强烈。

卡恩：非常对！

李：光线和天窗的尺寸、形式和深度都控制着整体的氛围，我在想这里是否有一个原则。设计和表达一个清晰的理念，仅仅只需要一种元素。这是很重要的，你不应该同时表达很多理念。

卡恩：每个项目都只有一个理念。

* 这里所强调的尺度是指以人为本的空间尺度问题，不同层级的空间尺度问题。

KAAN: To society and to human beings, for the actual people who work in the building and use it. In a way, we have to be humble, think about normal things, everyday things. People use the building, and the building is part of society. We work a lot with the topic of public and private, inside the building and in relation to the city. Also, with scale and abstraction, and different levels of abstract and tangible, hard, and clear.

LI: I understand, thank you. The third building I want to discuss is the Education Centre in Erasmus Medical Complex. I have visited it two times in the last two months. I like it a lot. I think in general space is defined by the composition of floor, wall, and ceiling. But in your education centre, space is controlled by the ceiling, the skylight. I think the whole atmosphere comes from it, it's very strong.

KAAN: Correct!

LI: The size of the skylight, the form and depth are all controlled. I wonder if that's a rule. To express a clear idea, there should only be one element. It is very important, there should not be several ideas.

KAAN: Every project – just one idea.

李：有些年轻的建筑师，特别是学生，总是想在一个方案中表达很多种理念，把方案变得很复杂，从你的教育中心中，我看到一个非常清晰的理念，非常强烈，一个建筑表达的越是简单，这个建筑带给人的感觉越强烈。

卡恩：是的。

李：还有一个问题，那就是教育中心大堂让我想起另一个人——凡·艾克，他在海牙的西部设计了一个天主教教堂。我去过那里，那里天窗对空间有同样强烈的控制，你的建筑和它之间是否有一些联系呢？

卡恩：没有，我认为它们之间没有联系。同时我不想给空间一种神秘感或者类似的感觉，它没有宗教含义，单对于我们来说内外转换空间很重要。这里原来是两栋楼之间的室外天台，我们的构思是希望把它变成室内空间。但是我们不想让屋顶仅仅是覆盖外部的空间，而是想把外部的空间变成一个房间。不是单纯地用一个顶去盖住它，让很多的光线进来，让人感受到是在两个建筑之间，这是一种空间转换。因此你感觉像是在一个房间里而不是在两个建筑之间，屋顶建立了一个连接，这是很重要的。同样重要的是，我们用一个元素去创造一个屋顶，把它设计得像是一个房间的顶棚，而不是一个在两个建筑之间的轻薄屋盖，使它看起来像是一个教堂的感觉。我时常觉得当我在这个空间之中时，

LI: Young architects, especially students, want to put many ideas in the design and make it very complex. And from the Education Centre, you get the sense of the idea as very clear, very strong. The simpler the building, the stronger the feeling.

KAAN: Yes.

LI: The great hall of the Education Centre reminds me of Van Eyck, he made a church in the west of The Hague. I went inside, it also has a skylight that controls the space strongly. Is there a relationship between the two buildings?

KAAN: No, I think there is no reference to that. We didn't want to make the space mystical or something like that, there are no religious connotations to it, but it was very important for us to invert the space. It used to be two buildings and an outside space between them, we didn't just want to make a covered outside space, but we wanted to turn it into a room. Not just cover it, let a lot of light in and make you feel like you are in between two buildings, but invert it. So instead of being in between two buildings, you feel like you're in a room, the roof was so important to establish that connection. It's also important that we use one element to create a roof that looks like a room ceiling, not a thin light roof on the top of two buildings, it is almost like a cathedral or a church. I often feel when I'm in that space, if I close my eyes, and open them again - instead of seeing the space, I see I'm in between two

我闭上双眼，再睁开，我不是在看这个空间，而是看到我在两个建筑之间。然后我闭上双眼再睁开，我又是在这个空间里。你明白我的意思吗，会很难懂吗？

李：可能是有一点难懂。

卡恩：我认为同一个东西可以给人不同的感受是很有趣的。你可以感受到这是一个在两栋楼之间被屋顶覆盖的空间，同时也可以感受到这是一个房间。（伸手拿纸）

李：你可以使用它。

卡恩：（在纸上画着一些东西）你有一个建筑，还有一个建筑，你把一个屋顶放在上面，通过那个屋顶你创造了这个空间，有个人站在这里，可以感觉到那是一个被放在两个建筑物之间的屋顶。我们在两个建筑之间加一个屋顶，用这个屋顶将外部空间遮盖，然后我们强化这个空间给人的感觉，将屋顶和建筑之间的关系变得相互关联，使这种感知变得更加强烈。这么做之后，这些强烈的要素开始发挥作用了，创造出了一个新空间，这就是我们想做的。这不单单是一个空间，而且是一个房间。

李：另一个问题是关于你对材料的使用。这里有一个放置书的墙，如果我离远看，它看起来都是白色的，我认为这是非常抽象的，我已经不能辨明其结构特征和材料特征了，我认为这是非常抽象的表达。

buildings. And then I close my eyes and open them again, I'm in that space again. Can you understand what I mean? It's a bit difficult.

LI: Maybe a little difficult.

KAAN: I like it when something can be experienced in different ways. You can experience it like a space covered by a roof, but you can also experience it as a room. (reaches for the paper)

LI: You can use it.

KAAN:(Drawing something on the paper) You have a building, and another building, you put a roof on top of it, and with that roof, you create this space. If somebody stands here, they can feel it as a roof that's put on these two buildings. But then, this space down here, it remains a covered outside space, but by making this very strong, and very much integrated with this, it becomes stronger, and this is behind it. So, this thing starts to work and creates a new space, and that's what we tried to do. It's not a square but a room.

LI: I have another question about your use of material. The library wall, looking at it from a distance, I saw it as all white, very abstract. I couldn't distinguish its structural or material features, it seemed very abstract.

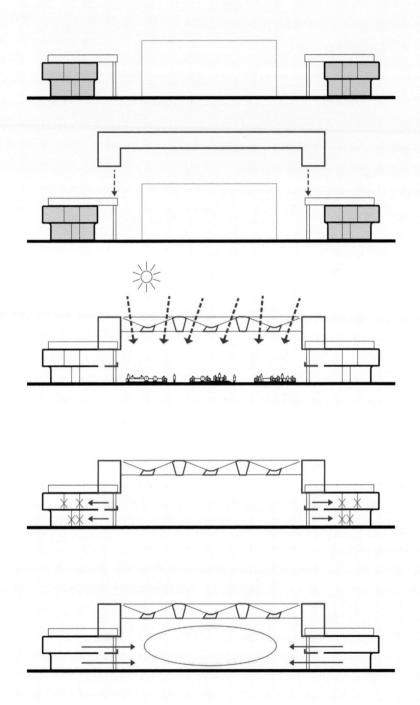

卡恩：目的就是创造这样的空间（指着手绘图），因为我们加入了所有现在的和新的结构，我们想把所有东西都合并，不希望看到新和旧之间有什么差别，所有事情都应该是新的，唯一的方法是让所有物体都变白（笑）。

李：好的，我明白这个道理。这和我想象的完全不同，没有我想的那么复杂（笑）。我想让你的想法被更多人明白。

卡恩：非常简单。我们的项目都是建立在简单、清晰的理念上的，然后通过感受去完成。你所做的事情是非常激进的，因此你必须始终如一和激进。[*] 这不是极简主义的，有时是简单的，但不是极简主义。我们总是有一个目的，我们不落俗套，做任何事情总有一个原因。

李：你的设计是非常清晰、合理的，但是如果我没有机会和你交流，我仅仅是看你设计的建筑无法知道背后的原因，不知道怎么产生出这样的结果，这对我来说是非常难的。因此我非常高兴能有机会跟你直接交流，我想把我们交流的内容传达给我的学生。

卡恩：当然可以。

[*] 这是在解释建筑师个人的设计观念和设计中的操作路径，建筑师在具体的设计过程中不会转换操作路径和方向。

KAAN: Yeah, the goal was to create this (points to the drawing), because we had all this existing structure and new structure, we wanted to merge everything into one, so we didn't want to see the difference between old and new, everything had to become new, and the only way was to make everything white (laughs).

LI: OK, I understand that. It is completely different from what I imagined, not as complex as I thought (laughs). That's why I'd like to introduce your ideas and let more people know.

KAAN: Very simple, our projects are always based on very simple, clear ideas, but then to get people to feel it. You have to be very radical, you must be very consistent and radical. Not simplistic, sometimes simple, but not simplistic. It always has to have a purpose - it's not banal, it's for a reason.

LI: Your design is very clear, very reasonable, but if I didn't have a chance to talk about it with you, I would just look at the buildings you designed and not know the reason behind it, I don't know how to produce such a result, it is very difficult for me. So I'm so glad to have the chance to talk directly with you and share it with my students.

KAAN: Of course.

李：不同的环境产生不同的结果。最后我想知道什么观念带给你灵感去创造这样一个结果？代替去追赶另一种方式，坚持深度的研究。是否有一种社会责任？简单来说，我想知道你会怎么去形容你的设计方法，你可以概括一下吗？什么影响了你的设计方法？

卡恩：我认为我们总是试图去找寻我们真正想要的是什么，这是很重要的事。每一个项目都有一个关键点，就是建筑和城市、社会之间的联系，我们试图去寻找为什么建筑在这个位置，这栋建筑存在的理由，这个建筑的目的，还有它对整个城市的意义。一旦我们定义了这个，我们就几乎定义了关于这个项目的所有想法，然后我们试图去接近主要的设计理念，我们采取非常激进的方式进行推进，它很简单，非常清晰，我们需要使它变得更强烈同时令人可以接受。我们需要在细节上投入很多的精力，我们必须去小心地对待它。这不是极简主义，而是简单的理念。简单的理念需要很精确的和谨慎的操作，需要很多耐心和关心。因此要做的是去发现，或者尝试去发现项目背后的主要问题，并且定义它。

李：你的意思是要发现关键点，关键的问题。

卡恩：是的，这个关键点驱动了整个项目的主导思想，同时非常接近它，但是我并不附加其他想法，而是接近它然后试图去尽可能地用激进的方式去深化。然后这个建筑就有了自己的性格，就像一个人一样有了自己的身份特征，这就

LI: Different environments produce different results. Finally, I want to know what concept inspired you to create such a result? Is it a social responsibility to insist on in-depth research instead of catching up?Lastly, I want to know how would you describe your design approach,Could you sum it up for me? What influences your approach?

KAAN: I think we always try to find one important, essential issue in each project, and it has to do with the relation of the project with the society and the city. We try to find the link between the purpose of the building, and its physical presence in the city. Once we define that - the overall idea of the object - we try to stay very close to it. We develop it radically, and because it is simple and clear, we need to let it become strong and at the same time, become acceptable. We need to pay a lot of attention to detail, we have to make it carefully. It is not minimalism, but it is simple ideas. Simple ideas have to be made very precisely and very carefully, with a lot of attention and care. So what we have to do is to discover, or try to find the main idea behind the project, and to define that.

LI: What you mean is the key, key question.

KAAN: Yes. Find the key drivers for the entire project, and then stay very close to that, not adding other ideas, stay close to that and try to be as radical as possible in development. That's when the buildings get a character, they become like people and

是为什么建筑都各不相同。就像最高法院、伊拉斯谟大学医学教育中心那个黑色的和白色的中庭，它们都是不一样的，但都有很强烈的内部一致性，因为它们都是基于强烈而简单的想法。

李：是的，你的意思是，建筑是一种与时尚、雕塑、绘画完全不一样的艺术。因为其中包含技术和应用的因素，所以怎样去保持一种激进的态度，而不是仅仅表达自己的想法，是完全不一样的方法。

卡恩：我认为我们建筑师设计的每一个项目，不应该仅仅是对使用者或者客户有所帮助，更应该为城市带来益处，对建筑所在的地方有积极的作用。环境必须变得更好，建筑必须产生积极的作用。这就是我们想做的，其实是很简单的。

李：我相信在我们交流之后，我的想法也会变得很简单。

卡恩：我认为采取一个谦虚的立场并试图与我们的背景找到共识是非常有荷兰特色的，这与我们的文化有关。如果你看过荷兰的古典绘画，我们是第一个画平民倒牛奶的国家，还有普通人的家庭生活，这些场景都被古典大师描绘出来。在比利时，他们还在画着上帝和有关信仰的主题时，在当时的荷兰，维米尔已经开始画普通人的生活了。我们必须对我们所在的社会做些什么，这些将简单地反映在我们的工作上。

have their own identity, that's why the buildings are so different. Just like the Supreme Court, Erasmus, the black court and white court, they are all very different, but there's a very strong internal coherence because they are all born out of a strong simple idea.

LI: Yes, so you mean architecture is a different kind of art than fashion, sculpture, drawing... because there are elements of technology and use. So, maintaining a radical attitude instead of just expressing your ideas are a completely different approach.

KAAN: I think every project we do, that all architects do, should not only facilitate the users or the clients, but it should make the city better, it has to contribute to the place where you put the building. The environment has to become better, the building has to contribute to it positively. That's what we want to do. It is very simple.

LI: I think after our talk, my ideas can also become very simple.

KAAN: I think it's also very Dutch, to take a modest position and try to find consensus with our context, this has to do with our culture. If you look at the classical Dutch paintings, we are the first country to draw people pouring milk, and normal people doing normal domestic things were painted by classical masters. While in Belgium they were still painting gods and religious things, in the Netherlands Vermeer was painting milkmaids. So, this has a lot to do with the kind of society we are a part of, that's simply reflected in the work.

李：你的意思是一个建筑的个性是被国家这个强烈背景影响的，国家强烈地影响着建筑师该怎么去建造和思考。我认为荷兰是一个很理性的国家，由于海拔非常低，地势呈现出一种十分水平的状态。如何去利用有限的资源发展科技来重新定义环境，这是很重要的。

卡恩：是的。生活在荷兰这个国家，我们总是要彻底重塑我们的环境，因为它本身并不适宜居住——它不安全，并不是极端恶劣，但算不上对人类友好。如果你拥有一小块土地，希望在上面建造一个房子，因为水的问题你需要和你的邻居还有其他人一起合作。由于荷兰的特殊地理原因，一旦你开始挖掘土地，你会发现有很多水涌出来。这意味着你不能自己一个人操作，你需要和其他人合作。很多个世纪以来，城市通过大家通力合作而形成，人们需要去一起工作，不是因为他们喜欢一起工作，而仅仅是因为他们必须这样做。所以我们有着一起工作的很强烈的必要性，但是同时也需要一些私密性。我认为这种博弈和平衡是文化中很重要的一部分。我们总是在寻求共识，不是坚持己见的，而是用一种大家都认可的方法去做事情。

李：是的，一开始你不得不这么做，最后这就成了你们文化的一种个性。这和中国是不一样的。

LI: So what you mean is that the personality of architecture has a strong background in society, nationality has a strong influence on how to do and how to think. I think the Netherlands is a very rational country, due to the very low altitude, the terrain presents a very level state. How to use the limited resources to develop science and technology is used to redefine the environment, which is very important.

KAAN: Yes. To live in this country, we have to shape our environment completely, because it's not liveable by itself. It's not safe, it's not extremely hostile, but it's also not very friendly. If you have a piece of land, and you want to build on it, you always have to work together with your neighbours and others, because of the condition of the water. Because of the special geography of Holland, if you start digging, you'll find water everywhere. That means you cannot operate on your own, you have to operate together. So already for centuries, the cities were built as collective efforts, people have to work together, not because they like to work together, but because they simply have to. So, there is a very strong relationship between the necessity to work together, but also need for privacy and individuality. And I think this game and balance are very important in our culture, we tend to always seek agreement and consensus without making things juridical but do it in a normal way.

LI: Yes, you have to do it at the beginning, but it becomes a personality of your culture in the end, which is different from China.

卡恩：是的，不同地方的方法是不一样的。在中国是不同的，在德国是不同的，在西班牙也是不同的。但是我们都是建筑师，我认为我们的社会是城市共享的，是城市创建的，我们的文化是在城市里产生的，而不是在农田中产生的。我们的进步和完整的演变是在城市里发展的，全世界都一样。如果你在短时间里观察城市，观察建筑风格，你会变得很迷惑，因为你只能看到流行的东西；但是假如你在很多的时间段观察它们，会发现城市演变是非常简单和清晰的，然后你会理解城市为什么会变成现在的样子。比如一座城市依山傍水，又比如一座城市有防卫需求，等等，每件事情都是可以解释的。假如你观察它，你总是会找到它背后的逻辑。一旦你了解了这个逻辑，我们就可以解决它。我认为这就是我们一直在寻找的。我们并不是在寻找建筑学的理论、现代主义或者类似的东西，我对这些一点都不感兴趣。我仅仅对事物的深层次逻辑，还有关于人的一切感兴趣。现代主义的思维，我的意思是，20世纪的现代主义思维是去把世界和问题分解成部分，然后为各部分找到解决方案。像是一个医生医治你的鼻子，另一个医生医治你的眼睛，等等。建筑学也是一样的道理，一个让人居住的建筑，我认为当代的文化是更加综合化的。我们试图找到比现代思维更为

KAAN: Yes. But I think everywhere is different. It is different in China, it is different in Germany, in Spain... But our job are architects, I think our societies are shared and shaped by cities, our culture is produced in cities, not in the farmland. Our progress and entire evolution are developed in cities all around the world. And if you look at the cities, if you look at architecture in a short period, you will have a lot of confusion because you see only fashion... But if you look at them in a longer period, things become very simple and clear. Then you understand why the cities became as they are... Because perhaps there was a river and there was a mountain, and you had to provide defence walls and so on, everything is very explainable. So, if you look at that, you always find a main driving logic behind it. Once you understand that logic, then you can work with it. And I think it is what we are constantly seeking. We are not seeking a theory of architecture, or modernism, or whatever, I'm not interested in that at all. I am only interested in the deeper logic of things, and the human aspect. Modern thinking, I mean the 20th-century modern thinking, tends to divide the world and problems into pieces and make solutions for those pieces. Like with the body - you have a doctor for your nose, a doctor for your eye, for everything. And the same in architecture, a building to live in... I think the current culture is more about integrating. We're trying to find more integrated complex solutions than modern thinking did. Modern thinking tried to simplify the

综合的复杂解决方案。现代思维单单是通过挖掘分离的每件事物来找到难题，为每一个难题提供一个答案。通过新的技术，我们可以再一次这样做，我们可以将更多的事情合并，同时找到解决更复杂的事情的方法。这仍然是很原始的，但比当时现代主义时期要先进得多了，我想这就是我们现在的处境。对于后现代社会，我不知道。我认为在全球化的社会，我们面临的挑战是去更加综合地思考，试图用一件事情解决十个问题，而不是相反，并尝试在不同的尺度上同时处理复杂性，包括对可持续性和一切事物的思考。我认为我们仍然是处在这件事的开头，这依然是非常原始的，但这也可以变得更好，在我们找到更多的解决方法之前仍然需要花费很多时间。

李：是的，就像你的设计，建筑的重点不仅仅是在功能、结构，或者好看的外表上，而是需要尊重使用者的感觉，这种感知是很复杂的。设计是为了解决这个问题，不是吗？

卡恩：没错。

李：非常感谢，我受益良多。

problem by taking everything apart and then making a solution for every individual aspect. And now, with new technology, we can do that again, we can combine things and find solutions for more complex issues. Still very primitive, but not as primitive as in modern society, so that's where we are, I guess. In post-modern society, I don't know. I think that in a global society, the challenge is to think more integrated, to try to solve ten problems with one thing, instead of the other way around. And to try to deal with the complexity, simultaneously on different scales, including the thinking on sustainability and everything. I think we are still at the beginning of that, it's still very primitive. We can become much better at it. It will take a lot of time before we can make more intelligent solutions to a problem.

LI: Yes, just like in your designs, the focus of architecture is not only on function, structure, or beauty but also on the feelings of users. This perception is very complicated. The design is the solution to this problem, isn't it?

KAAN: That's right.

LI: Thank you very much. I learned a lot!

对谈雅各布·凡·里斯（Jacob van Rijs）
& 娜莎莉·德·弗里斯（Nathalie de Vries）

访谈时间：2016 年 12 月 12 日

　　雅各布·凡·里斯和娜莎
莉·德·弗里斯是 MVRDV 建筑师事
务所的创始人，担任多所全球知名大
学的客座教授。MVRDV 事务所是当
今最有影响力的建筑师事务所之一。
MVRDV 非常关注荷兰整体的社会发展
趋势，不论在建筑或城市设计中，还
是在景观设计中，他们都希望表达一
种对社会生活状态的独有理解和关怀。

李：我想谈的第一个话题是关于 2000 年汉诺威世博会的荷兰馆。几年前，我在杂志上看过关于荷兰馆的资料，这个项目给我留下了很深的印象。我非常喜欢 2000 年的汉诺威世博会荷兰馆的设计，它是一个成功的、新的概念设计。在这个建筑里，自然环境是创造出来的。在荷兰如今有一种迹象，技术和自然不再分离。这让我想到了我在荷兰的研究，不仅是关于建筑，还涉及荷兰人对大自然的理解和态度。根据我的观察，一方面是人们保护森林和湖泊，并且很好地保护和保持原始的自然状态；另一方面是自然已经适应了人类的需要。这里有一个例子，在去吕伐登（ Leeuwarden ）的火车上，我看到一条河流从高速公路上流过。这让我非常震惊和惊讶，我认为这是荷兰人对自然的态度。保持自然系统良好的循环状态并介入它。因此，我的第一个问题是：荷兰馆的设计理念是否是这样被构思出来，并且与大背景有着密切的关系。你能谈谈设计的出发点吗？

LI: The first part topic is about Dutch pavilion at the 2000 world expo. Many years ago, I saw the material about the Dutch pavilion in magazine, the project left a very deep impression on me. I really like the design of the Dutch pavilion, it is a successful conceptual design. In this building, the land could be manufactured and the natural could be superposed. According to the situation in the Netherlands, there is a hint, technology and natural could no longer be separated today. And this reminds me of my research in Netherlands, not just about architecture but the Dutch understanding and attitude about the nature. According to my observation, on the one hand people protect the forest and lake to keep the original natural states; On the other hand, the natural has adapted to human needs. There is an example, on my train to Leeuwarden, I saw a river flowing over the highway. That gave me a strong shock, it surprised me. I think that would be the attitude of Dutch to the nature. It's the nature system that is involved while preserving is good circulatory state. Therefore, my first question is whether the design concept of Holland Pavilion is conceived in this way, and the related background. Could you talk about the starting point of design?

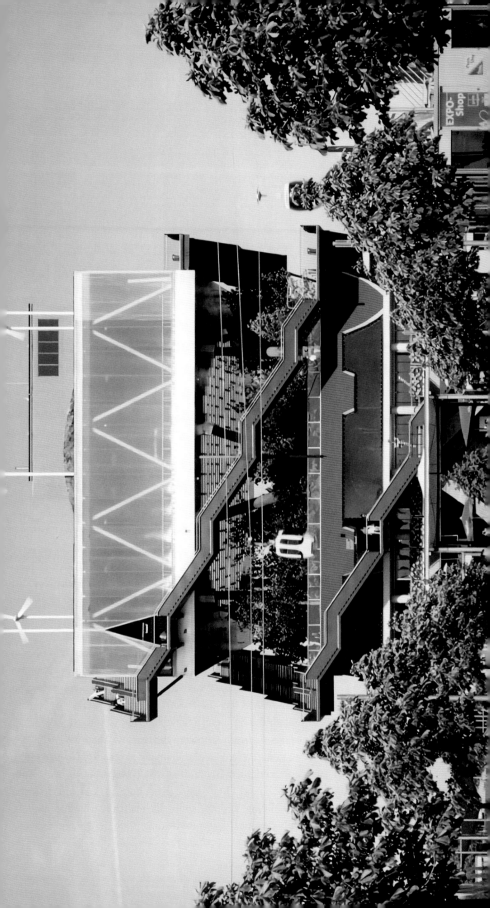

娜莎莉：是的，我明白这个问题，我先开始吧。你说得很对，我们被要求在世博会上展示荷兰的代表性。我们马上就看出这很有可能是一个展示荷兰未来发展的机会，但我们也必须把荷兰的土地和荷兰的特性非常强地联系起来，因为建筑总是受政府委托的。所以有趣的是，如何在德国展示一座有荷兰代表性的建筑物是非常重要的。第三个因素也是必需的，政府说他们想要一个非常成功的展馆，能吸引很多游客，它也必须包含一些精彩的景象。因此，我们所做的就是实现在那之前的几年里就讨论过要把强度和功能结合在一起的想法。荷兰政府在德国世博会预留了非常大的一块场地，我们根据场地的边界来切割这个体量，你也可以在公共场合看到这种切割，并真切地感知到场馆的面积，在这种边界已经限定的情况下，整个展馆不同的功能被堆叠，空间朝竖直方向发展展现了密度。我们认为，荷兰未来将成为一个人口稠密的国家，我们也知道这种情况正在世界各地发生，那么荷兰的做法将会是特别的，我们不仅建造建筑物，而且做各种各样的高密度农业，我们第一次把这些东西合并到一个建筑中。事实上我们已经有了一些参考，我们知道在荷兰有很多垂直农业，也知道从不

NATHALIE: Yes, we understand the question, I can start. Yeah, you are right, we were asked to make it, of course, a representation of the Netherlands, at the expo. And we immediately saw it is a possibility to make a manifest about the future development of the Netherlands as well, but we also had to connect the Dutch lands to the Dutch identity very strongly, because the building is always commissioned from the government. So what's interesting is, it's very important to show a typical Holland building in Germany. And the third element is also necessary, the government said they want a very successful pavilion with a lot of visitors, it also has to contain some spectacle as well. Yeah, so what we did is that we applied actually our ideas that we had communicated also in years before about intensity and function mixing to this place. The Dutch government secure a very large plot on the German expo, and what we did is that we cut it up, you can also see that in public occasions and feel the area of the venue clearly. In the case that the boundary has been limited, the different functions of the whole exhibition hall are stacked, and the space develops vertically, showing the density. Then we said in the future of the Netherlands, we're going to be a densely populated country, we also know it is happening all over the world, and what Dutch does would be particular, we do not only construct buildings, and do all kinds of high dense agriculture, we could also for the first time merge this things in a building, and in fact we have already a little examples, but we also know there was a lot of vertical agriculture going on in the Netherlands, we

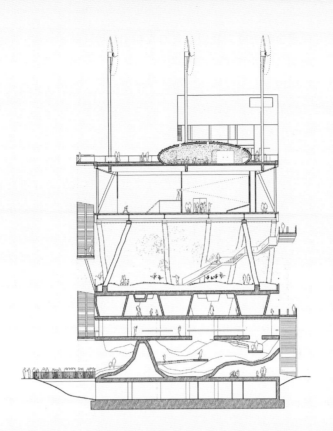

同的角度思考能量会很有趣。我们也对回收利用的做法很感兴趣，例如，对水的回收利用，我们认为这种建筑也很好地将自然引入城市，而且可以考虑城市和景观的结合，所以实际上这个建筑包含了很多信息。我们认为荷兰并不完全是一个大都市，我们不再有纯粹的景观，它总是有点混合。所有这些东西都聚集在展厅里，我想让人们可以去构想……

雅各布：荷兰馆的主题是"荷兰创造空间"，我们可以消费，我们可以开垦土地。

娜莎莉：现在我们开始在垂直的方向上进行建造，这样我们可以进行更高质量的致密化和优化，因为如果你不做任何事情的话，一半的荷兰将会在水里。事实上，这种制造密度的新方法就是我们想要的。

雅各布：这是在调试密度和空间之间的关系。这是一个故事，它可以通过一种社区的方式来完成，你可以说它是一种简单的方式，但是建筑是一种复杂的方式，它意味着任何人都可以享受它，然后人们意识到，哦，原来它是这样的，不是吗。这是一种沟通式的建筑，建筑从根本上说是一种信息，而不是展览。我们不希望它是一个里面只有展览的展厅，但最后我们也不得不砍掉一些东西，像剪辑一部官方电影那样。

also knew it would already be interesting to thinking about energy in a different way. We were also interested in the idea of recycling, for example by catching water, and using all this, we thought this kind of building is also perfect to bring natural in cities, and to think more hybrid about city and landscape, so there are a lot of messages actually. We already conceived that the Netherlands are not totally metropolitan and haven't clear empty landscape any more, it's always a bit of hybrid. All those things came together in the pavilion, I want to show that you can construct.

JACOB: The slogan is Holland create space, we can expend, we can make land.

NATHALIE: Now we can start to make lands vertically, by that we can make higher quality identification and optimization, because half of the Netherlands will be under the water if you don't do anything right. Indeed this new way of making density is what we want to bring together.

JACOB: So make emptiness and dense. So this is a story, and also it can be done in a way, community's way, you can talk it is a simple way, but the building is a complex way, it means that anybody can enjoy it, and later on people realize oh yeah it's like this, isn't it? It's a very communicative building, which is a message, not an exhibition, we didn't want it to be a pavilion which is a hall with an exhibition inside, in the end we had to ax a few things, like a movie in official way.

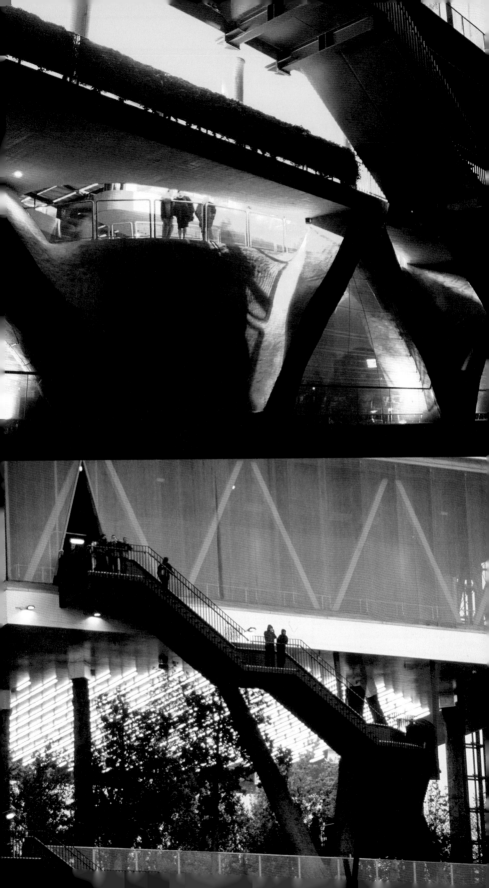

娜莎莉：入口还有一个主题公园，播放着节目，（游客）可以选择在主题公园处等待。我们也想在这里放进动物，并真正地种上蔬菜，但是荷兰政府不认可，我们便布置以鲜花。我们想要收集雨水，把它引入到大楼里，但他们也不认同，认为这太危险了，有人可能会生病。我们设计了雨水槽，但实际上也没有作用，因为它不是为整栋建筑而工作的。这是最富有流动性的一座建筑。

雅各布：他们很关心这个建筑是不是和我们展示的图片一模一样，这是我们向人们展示的东西，他们希望我们能做到这一点。这期间也会有一些调整，设计会有一些改变，有些做法被否定了，因为那些不是为了这个设计主题而做的，所以我们不能改变它的外观。

娜莎莉：现在看来，我认为那栋建筑是很有意义的，因为我们建造的这个建筑，是第一个展示出如何把能源和自然融合进建筑的建筑；另外，还可以循环利用一些资源，这是一座建筑，而不是景观。我认为很多被认为是进化的东西，现在正在发生。我认为这是一种征兆，对未来的一种预示，这很好，因为这是一次爆发。就像巴克明斯特·富勒（Buckminster Fuller），想象一下那个人造的球壳（1967 年加拿大蒙特利尔世界博览会美国馆），在某种程度上，他对客户的某些想法很了解。

NATHALIE: There is also a theme park in the entry playing shows, the theme park apply to waiting. We also want to put in animals for example, and really manipulate it vegetable, but the Dutch government says no it is too scary, in Germany we don't want to have any political problems, just put in flowers. We want to catch the rain water and move it to the building, then they say no it is too dangerous, someone may get ill. So we have the rain gutter outside but never actually work because it works not for the whole building. This is the most fluid building.

JACOB: So they are very concerned that it looks exactly like the images we made for the promotion, and it is what we show to the people, and they expect we're going to made it, there would be some changes, it also means to change the design a little bit, they were against because it is not for this design theme, so we can't change that looks.

NATHALIE: So I think look it now, the building really makes sense, because the first time we made the building, it was one of the first buildings that really show how you can put energy on and how you have trees and green higher level buildings, and you can recycle a lot of things, that can be a building rather than something landscaped. So I think a lot of things which were dent considered to be evolution, are now happening. I like to make it now a kind of announcement, a preview of the future, which is good because that is explosion. Like Buckminster Fuller, imaging this artificial bulbous dome. In a way, he knew the certain thinking of client well.

李：是的，你的设计让我觉得很惊喜，关于这栋建筑，即使放在今天，我认为它也是一个非常前卫的设计。

雅各布：显然这个建筑让德国认识了荷兰的特性。这座建筑现在是一座废墟，因为他们想要普通的建筑，他们不想这么做，这是一个复杂的建筑，它也不需要你再去回收利用了。你应该拥有所有权，这种所有权可以捐赠给一个人来做任何事情。*

娜莎莉：建筑的顶部是可拆除的，它是钢的，底部是混凝土，所以建筑的底部应该是大堂。例如，风车和树可以被移除。

李：混凝土结构还在吗？

娜莎莉：所有东西都还在，除了风车。

李：如果我早知道这个信息，就去参观一下了，下次有机会就去看看。第二个问题也是关于荷兰馆的，它揭示了一个垂直发展的观点，同时，各种技术系统已显示出一体化的状态，包括结构系统、绿色技术系统、自然系统、交通系统等。而不同的人对这个设计有不同的态度。作为设计者，你希望外界如何理解这个作品？是作为一种可持续设计的过程？还是一种激进的风格吗？抑或两者皆可？

* 此处雅各布意在强调建筑的可持续性、再利用的可能性。

LI: Yes, your design gives me a surprise, about this building, even today, I think it is also a very radical design.

JACOB: The straight thing is that now there, it teaches Germany the characteristics of the Netherlands. The building is still standing as a ruin, on one hand they want normal buildings, they don't want to do it, it is a complicated building; And also it is no need for you to recycle. You should have ownership, the kind of ownership donate to someone to do anything with it.

NATHALIE: The top was demountable, it's steel, and the bottom is concrete, so the bottom of the building was supposed to be lobby. For example, the windmill was just demountable, the trees could be taken out.

LI: Is the concrete structure still there?

NATHALIE: Everything is still there, expect the windmill.

LI: OK, if I got this message earlier, I would go there and have a look, I still have the opportunity. The second question is about the Dutch pavilion, which reveals a view of vertical development, at the same time, a variety of technical system have shown an integrated state, including structure system, green technology system, nature system, traffic system, etc. And different people have different attitude to this design. As a designer, how do you want the outside world understand this work? A process of sustainable design, a radical style, or both?

雅各布：我认为两者都有，因为人们在某种程度上过于表面化，而且它也有一个内涵，其背后的意义是很重要的，所以了解其中的意义能帮助我们理解这个特别的建筑。

娜莎莉：这个案例非常重要。

李：是的，基本要素是关于如何处理与人和环境的关系。我认为这是基本的。

雅各布：使用我们已有的空间是很容易的，因为它很重要，这是一种填充，它不断地成为一种吸引人的空间，剩余的空间越来越少，创造空间是很重要的，你必须要把它们重叠起来而不是像往常一样水平放置。

李：我认为这种激进的风格只是结果，从一开始你就想象到这仅仅是坚持思考的结果了吗？

雅各布：我们的客户，也就是荷兰政府，认为我们讲述的是一个好故事，它也应该是一个有趣的话题。建筑应该以一种展览的形式存在，它只是意味着一种激进的方法，可以确定这些在将来也会像现这样发生着，这是一种不同的氛围。

JACOB: I think it is both, because people was somehow spectacle, and it is also has a content as well, the idea behind is very important, so to get the meaning across it helps us to understand particular building.

NATHALIE: And this case is very important.

LI: Yes, the basic element is about how to deal with the relationship with the people and the environment. I think it is basic.

JACOB: It is easy to use the space because it is very important. That is populated, it keeps becoming one big a kind of attractive space, there is no empty any more, it is important to make empty, the things are put on top of each other rather than normally next each other.

LI: I think the radical style is just the result, from the beginning do you imagine the result just following the persistence of thinking?

JACOB: The client which is the Dutch government thinks we explain a good story, it should be an interesting topic and the building should exist as an exhibition, and it just means a radical step, much sure it will be happening this today, it is a different atmosphere.

娜莎莉: 但我认为，对我们来说，需要提高生活质量，提高密度，我们用合理的方式进行交流，可以是技术，可以是景观，也可以是城市设计。这种方式在城市设计与建筑之间是非常重要的——系统的稳定性和技术是工具。我们也看到了快速的发展，我们看到的保存下来的艺术，现在已经过时了。我们相信什么，我们就必须发展我们如何建造建筑、城市和社会的想法，我们至少可以借助于技术，我们不应该放弃它。而且这也不是我们的目标，因为技术会改变时代，一个好的建筑应该永远存在。技术使以前不可能发生的事情变得可能，但确实有工具，它是我们的工具。有时我们做的建筑物还有些实施性，我们总是用"模范"这个词，有时我们做了原型和声明，有时因为我们是第一个这样做的人，这让人对建筑有很高的期待。所以有时更多的是你要让客户愿意突破边界，这是我们的任务之一。我们应该有一个想法，一个关于城市哲学的想法。我们的哲学是如何创造一个更好的高密度环境，如何让人们生活在一起，如何创造更加舒适的功能区域，这就是我们的方法。每一个项目都有不同的混合，对可能性的倾向，时机、资金、科技，所有的事情都应该是平衡的。

NATHALIE: But I think, for us it is always about making a good quality of life, of a higher density, we communicate with each other in possible means, also technology, also landscape, also urban design, it's high great between urban design and architecture, and system stability, technique, they are tools. And we also see a rapid development forward, the saved art we saw is nowadays already, old fashion. We believe that we have to develop our thinking about how we could build the buildings, and cities and societies, at least technique helps us to make it, we shouldn't be abandoned on it. And it is also not our aim to go, because technology would change the age, a good building should always survive. Technology makes things possible now, which may not possible before, but there is really instrument, it is instrument for us. Sometimes we make buildings that are still a bit implemented, we always use the word "exemplary", and we know sometimes we made prototype or statement, sometimes because we are the first to do it, which also pays high expectation for a building. So sometimes you get clients willing to push the boundary, it's more, it is one of our tasks. We like to think we have an idea, we have city philosophy. Our philosophy is about how to create better dense environment, how to make people live together, how to make functions make more agreeable area, this is our path. Each project has different mixture, the bending on the possibilities, occasion, finance, technology, all the things should be balanced.

李：根据你所说的，基本元素在设计中是重要的，不仅仅是遵循某种风格，风格只是一个结果，因为思考的自然过程就是结果。我认为在代尔夫特理工大学，建筑教育也遵循这样的方式，不只是去看漂亮的形式，我认为分析元素是重要的。

娜莎莉：你认为教育很重要，是影响我们思考方式的一部分。我学会了从本质的角度去思考。

雅各布：去解释这个故事。

娜莎莉：是的，清晰的概念，整合，或者从各种方面。一个开放的设计包含很多，所有的这些，建筑技术、社会因素，要去创造一个良好的社会。我们也说这是非常重要的部分，也许历史因素较少一些。

雅各布：这是关于荷兰社会的，关于概念的，你必须说服人们这是个好想法。如果你说，这样设计是因为我喜欢它，因为它美丽，很难仅仅因为说它是美丽的而说服别人，这是没有意义的。例如，我最喜欢的颜色是红色，我最喜欢的颜色是蓝色。

娜莎莉：或者说在这个环境中它应该是黄色的。

LI: So, I think according to what you said, the basic element is important in the design, not just to follow some kinds of style, the style is just a result. As nature process of thinking is just result. I think in Delft, architecture education also is following this way, not just go head to the form that looks beautiful, I think analysis the element is important.

NATHALIE: You think that education is important, is a part that influences the way we think. I have learned to think in an internal way.

JACOB: To explain the story.

NATHALIE: Yes, clear concepts, integration, or of all kinds of aspects. An opening design contains all this things, building technology, social aspect as well, to greet good society, we also say this is absolutely important parts, maybe there is a little bit less history.

JACOB: That's always about Dutch society, about concepts, so you have to convince people about this good idea. If you say, well, the design is like this because I like it, because it is beautiful, it is very difficult to convince just by saying it is beautiful, it doesn't make sense. You may think it is beautiful, but somebody thinks it is not beautiful, my favorite color is red, my favorite color is blue.

NATHALIE: Or say in this context it should be yellow.

雅各布：如果你带着人们的思维走，他们说同意，这样的结果有可能会非常极端，他们认同论证，这也是可以理解的。人们往往认同一个结果。但也有可能，这个结果会非常难以预料。

李：我认为可能是分析、解释也很重要，如何理解他人。

娜莎莉：我认为代尔夫特理工大学和我访问过的国外大学是不同的，其他国家的一些大学总是过于强调科学，在代尔夫特我们总是对设计研究、文化联系、社会联系感兴趣。大学设计的社会组成部分是非常重要的。

李：我在荷兰也采访了其他建筑师，我想也许这是荷兰建筑师的共同特征。不像是一些建筑师在任何方面都非常强调外观，我认为思考的侧重点是非常重要的。接下来的问题是，从我的观察来看，荷兰馆就像一个集合体，它讨论了人、自然和科技的深层关系。就像勒·柯布西耶的萨伏伊别墅，也是一种宣言，关于如何处理人与人之间的关系、功能和技术。但那个时代已经过去，现在如何实施可持续发展都是不同的，有不同的步骤。荷兰馆的概念是否影响了你后来的作品？13年过去了，这种观念是否改变了呢？你能给我举个例子吗？

JACOB:If you get people along the way, and they say "I agree". The result can be very extreme, they accept the argumentation, it's understandable. People accept the result, which is not, maybe, they can be unexpected if they agree with the argumentation they accept.

LI: I think maybe it is analysis, and the explanation that is also important, how to gain understanding of others.

NATHALIE: I think it is different between Delft University and the foreign university I have visited, in other countries they always put emphasis too much on science, and in Delft we are always interested in design research, culture contact, and social contact. The social components of the design in university is very important.

LI: I think maybe, I have the interview with other architects in Netherlands, I think maybe it is the common specialty in Netherlands. It's not just other architects put much emphasis on the facade in any other aspects, I think the emphasis is very important. And the next question is that, from my observation the Dutch Pavilion is like a declaration which discusses the deep relationship of people, nature and technology. As the Le Corbusier's villa Savoy, which is also a decoration, about how to deal with the relationship of people, function and technology. But the time is passed by, now how to put sustainable development is all different, with different step. Whether the conception of Dutch Pavilion have been affecting your later works? And 13 years passed by, whether the conception has been changed? Could you give me an example to this part?

雅各布：它没有改变，我们仍然有这个故事，我们必须以有效的方式使用所拥有的土地、空间。我们可以结合功能，使它更令人兴奋，更具有吸引力。世界，可以变得更紧凑来使大自然更加惊奇，这还在继续，并持续很长一段时间，它仍然可以很有效，甚至更有效。城市的密度正在增加，紧凑的城市运动正在进行。在同一道理上，它需要一种解决方案，在这个过程中，态度并没有改变，因为我们学会了以不同的方式工作，我们有更多的经验。

娜莎莉：我们老了。

雅各布：但我们仍然想要保持这种态度，当然是以更好的方式。我们现在已经做了，这个想法可能仍然是一样的，但看起来可能不同，我认为这个概念仍然很强烈。

娜莎莉：我想我们现在已经能够使用两三种元素，每次在一个建筑上会实践一个想法。因此你会看到在一个项目总是使用一种屋顶，我们总是试图在建筑物中获得绿色元素，我们总是试图在建筑物中获得公共空间。

雅各布：外面的楼梯出现在岩石文化大厅*里，我们在那里组织了盘旋的楼梯，我们现在纽约的一个项目中也有这样的楼梯。

* 这里说的是 2000 年世博会荷兰馆中的一个空间。

JACOB: It didn't change, we still have this story that we have to use the land, the space we have in efficient way, we can combine functions, to make it more exciting and more attractive. The world can make more exciting nature by being more compact, this still continues, stays for a long time, and it can still be very valid, maybe even more valid. The densification of the cities is expanding and expanding, the compact city movement is ready. So in the same line, it tries to make solutions. So in the way, attitude has been not changed, because we learn to work in a different way, we have more experiences.

NATHALIE: We are older.

JACOB: But we still like to try to keep the attitude to same, but of course in a better way. The idea might still be the same, but looks maybe different, but I think the concept is still very strong.

NATHALIE: I think we have been able to use two or three elements, each at time from the idea on the building. So you will see a project always uses roofs, we always try to get green elements in buildings, we always try to get public spaces in buildings.

JACOB: The staircase on the outside is also in the rock culture hall, where staircases are spiraled in the around, we have now a project in New York where we also have staircases like this.

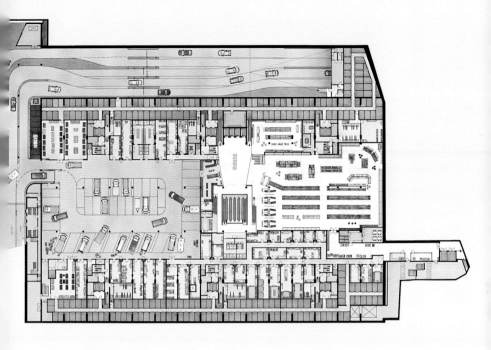

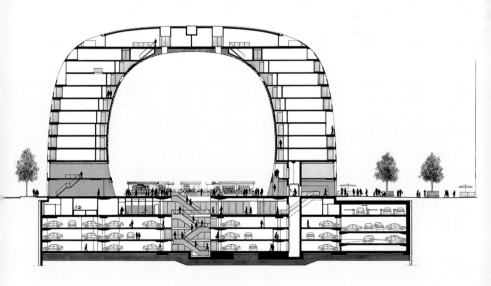

娜莎莉：我们试着让城市的公共空间融入建筑，在（鹿特丹）市场大厅中你可以看到这种想法。但我认为，因为我们通常要整合两三个方面，我们需要仔细地设计。这个建筑可以变得更加有混合性，可以结合一些新的类型，或者混合的类型，实际上我们可以做得更明确。所以在我们的展馆里，目前实施方案还是不够的。我们还说，你可以把一些东西拿出来，再把其他一些东西放进去。之后，我们当然可以更精确地了解你想要怎样做。

雅各布：我想应该考虑把工作坊的下一个研究方向定为展馆，该怎么做，现在该做什么，下一步是什么。

娜莎莉：我想在第一时间，人们也明白了我们的意思，我们正在把这个三维空间混合，现在我们真的做到了。我们将在荷兰做得更好。我们在一个正在发展的城市工作，我们现在可以将我们的想法实现出来。更有趣的是下一步，下一步可能来自于技术，因为我们正处于我们所知的建筑的边缘，我们必须要改变。

雅各布：改变整个能源革命，我们将产生巨大的影响。

NATHALIE: We try to bring cities up and down, into the building, and even in the market hall, you can see. But I think because we usually integrate two or three aspects, we need work carefully to make the design. The building can become a bit more hybrid, and can develop new typologies, or mixed typologies with that, so we can be more precise actually. So it's less implemented in our pavilion. We also say you can take something out and put something else in. Afterwards, we are of course able to be a bit more precise about how you want to make it.

JACOB: I think I should consider to do design studio next method on a pavilion: what to do with it, what to do now, what's next.

NATHALIE: I think for the first time, people also understood what we meant, we were making this three dimensional mixed, and now we are really doing that. We are going up in the Netherlands, we are working in growing cities, so we can apply now. What maybe more interesting is the next step, it probably does come from technology, because we are at the edge of where we know are building, with building we have to change.

JACOB: Whole energy revolution, we will have a big impact.

娜莎莉：还有关于回收使用和封闭循环的问题。认为建筑是材料的储存仓库，并且你可以将材料从仓库中分离出来的观念、我们与材料的不同联系方式、我们做出的选择和能源的消耗，以上几点将影响我们未来几年的建造方式。同时我们也意识到在设计技巧方面的工作越来越不一样，因此不同的建筑顾问之间的知识交流也在迅速变化，设计中的不同部分之间的关系也在发生变化，这在未来几年将非常重要，我认为这将会产生一些影响。目前我们已经见证了在这个过程中设计的方式正在改变，作为建筑师，我们可以再次承担更多的责任，就像有人说的那样，把所有这些想法结合在一起。在荷兰，致密化也将是我们的选择，你也看到了汽车的消失。例如，在一些地区，对可再生能源的兴趣，对不同的材料选择，我们依然对很多东西都不了解，关于我们怎么利用这些，现在还在试验中。

雅各布：产业结构将会发生很大的改变，因为现在离让它实施起来还有很大的差距。你知道我在办公室里工作的方式，作为一个团队，我、工程师和其他人一起工作，Revit 和 BIM 对我们如何建造有很大的影响。因为这是一步一步的，我认为整个建设的心态将会是，它来自双方，不仅来自建筑师，而且来自合作伙伴。

NATHALIE: And also the recycle and the close circles. This ideas that buildings are just a storage of materials and you will be able to disconnect from the depot of material, the different relation we will have with material, choices that we will make, and also the energy consumption will affect on how we will build in the coming years. And also we realize that we work more differently in design tips, so the knowledge exchange between different consultants of construction, is also changing rapidly, and relationship between different parts in the design is changing, so this will be so important in the coming years, I think this will affect. For the moment we already witness in the way we make our design in the process, they are changing, we can take more responsibilities again as architect as one said bring all these ideas together. Also the densification will be our choices, also in Netherlands, and you see the disappearance of cars. For example, in some areas, being too interested in renewable in energy, different selection of material, but we are still fresh of everything, how we are going to use all this are experiment.

JACOB: Industry construction extreme will change, cause right now there is still a big gap between making its work. You see the way I am working in the office, that I work as a team with engineer and others, Revit and BIM have big impact on how we can build things. For it is step by step, I think the whole mentality of construction will be, it comes from two sides, not only from the architect but also from the companies.

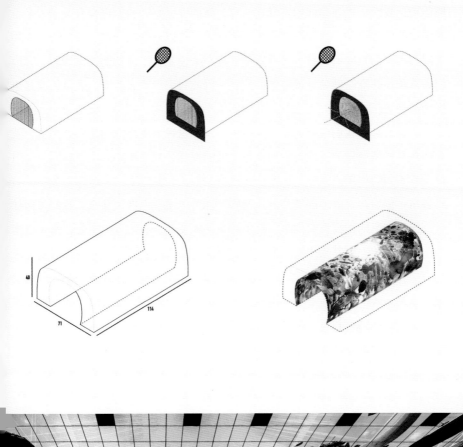

娜莎莉：我们可以在房间里发现这一点，例如办公室，我们现在有关于未来不同用途的更多讨论，我们试着建造更多在未来可以改变的建筑。我们用更高的上限和不同的角度看待技术，这对商业客户来说是一个复杂的探讨话题。客户们通常关注住宅、办公室或零售的问题，因此我们很难去改变、去重组或建造多功能的建筑。但我们在尝试，并取得了小小的成功，同样，就像每一个在转型中的建筑师，我们应该有很大的自由，我们也越来越多地创造混合项目，就像你在某些城镇做的，或者我们在丹麦做的那样，越来越多的建筑在同一个屋顶下有不同的作用。

雅各布：仍然存在着不同类型的建筑师，你可以说我们擅长什么，我们想做什么，但我们不能做完所有的事。这个行业如此宽泛，有太多的主题，你可以提出一个具体的主题并集中注意力，但如果关注过多的主题，你永远不能集中注意力。我们的工作室不断壮大，保持团队力，专注于我们正在做的事是最重要的，否则，MVRDV 会变得不像那个唯一的 MVRDV。因此我们试着把焦点集中在我们应该坚持的事情上来，我们必须把它发展到一起来思考，这个项目需要我们在这里做什么。

NATHALIE: We see it in the house, for example, the offices, we have more discussions now, about different uses in future, so we try to make more buildings that can transform in the future. We have higher ceiling and different ways to look at technique, it is a complicate discussion with commercial client, they usually focus on housing or offices or retail, so that's very difficult for us to change, to remix or make transformable buildings. But we are trying, we have a small success, and the same, of course like every architect evolves in transformation now, we should actually be of a large of freedom. And we create more and more hybrid projects just like what you do in some town or what we do in Denmark, more and more buildings have very different functions on the one roof.

JACOB:There are still different types of architects, so you can say what are we good at, what we want to do, we cannot do everything, the profession so broad, so wide, so you can come a sort of specific theme and focus on the theme. Too many themes, you can never focus. As the offices have to, because we are now much bigger, keep to group, focus on what we are doing, this is the most important. Otherwise, MVRDV becomes like… there are many MVRDVs coming, which might become different, so we try to keep the focus on what we should stick to, we have to develop it together to think about it, the project should advise what we try to do here.

娜莎莉：例如住宅，我们仍然想尝试将开放式设计和建筑创造多样化结合起来，同时，也要考虑人们自身如何影响他们的生活方式，因此，更多地关注想象力是过程的一部分。但公共建筑变得更加混合，因为许多公共建筑现在以某种方式被私人资助或者是混合地重新调整，它们不再是纯粹的公共建筑了。所以我们需要创建新的程序，使它们混合在一起。即使是市场大厅这样的商业建筑，有趣的是也可以发现公共区域，它们也可以作为一个大厅服务于城市。这些是我认为我们正在研究的内容，当然，我们也在一些技术领域，开发新的建造方法和规则。

李：还有一个问题是，我发现一些建筑师在同一时期提出了"人造景观"的概念，比如赫尔佐格与德梅隆和妹岛和世。同一个词背后的设计内涵有什么区别？我想也许前两个人更关注建筑与人类感知之间的关系，但你的设计可能与可持续发展有密切的关系，是这样的吗？

雅各布：在很多地方，这也是一种非常人性化的方法。凡·艾克在学校里有很大的影响力，他仍然在代尔夫特。我们也曾预言过这一点，这意味着这种人性化的方法仍然是帮助人们相互联系的重要角色，建筑师如何才能改善人们的生活。

NATHALIE: Housing for example, we still want to try to make the mix between open design and architecture create diversity, and at the same time, we also think about how people can influence the way they are living by themselves, so more attention for imagination is a part of the process. Public buildings become more hybrid, specially because a lot of public buildings are now somehow privately funded or in mixed retune, they are hardly any pure pubic building anymore. So we need to create new programs, mix for that. Even commercial buildings like the Market hall there is interesting to figure out that can have public compartment, they can also serve the city as a hall. So these are developments I think we are working on, and of course we also develop new ways of construction and rules in some physics.

LI: Maybe a very small question there is, I think I find several architects raising the conception of "artificial landscape" in the same period, such as the Herzog and de Mellor and Kazuyo Sejima. I wonder what the difference is in the design connotation behind the same word. I think maybe the first two is pay more attention to the constructing relationship between the building and human perception, but yours maybe have close relationship with sustainable development, is that so?

JACOB: That's also very humanistic approach in many places. A. Van Eyck, he has much influence in school and he is still in Delft. We have predicted that as well, so this means this element is still an important role to help people relate to each other, how architect can improve people's life.

娜莎莉：这也是我们在建筑书籍上关注的东西，展示人们的使用情况，倾听他们对建筑的看法。（李：使用它们的感觉。）这是非常重要的。

李：是的，我喜欢它，就像一个月前我在 BK 买的那本一样。

娜莎莉：所以建筑更像是它的用途。

雅各布：这里更多的是关于人的事情。

李：人们对感知的感觉，我认为这是需要回应的一个非常重要的事情，这应该影响到设计。第二部分是住宅。你们设计了许多住宅项目。在相同的项目中，有很多种单元，例如 Solidam 和 Mirador。从我的观察来看，Solidam 的表面存在着各种各样的材料和颜色。它产生一种幻觉，任何一种材料的特性都被削弱了，与此同时，整体产生了复杂的感觉——一种破碎的感觉。你能阐明第一个设计吗？为什么要在同一个立面上设计不同的颜色和不同的材料，为什么会对内部不同的单元做出这样的回应。这很有趣，我想知道原因。

娜莎莉：我认为这与试图摆脱现代建筑的限制有关。当然，在代尔夫特我们已经认识到了在建筑设计上怎样是好看的建筑。住宅建设大国都会有很多限制。但我们知道我们必须继续建造大型住宅小区，这里有很多不同的部分，但也展

NATHALIE: That's also what we focus on the building's book, to show how people use, to hear ideas about the building, (LI: The feeling of how to use it.) that's very important.

LI: Yes, I like it, just like the one I bought in the BK a month ago.

NATHALIE: So building is more like its use.

JACOB: That's very much about people.

LI: How people feel about perception, I think it's very important to response, there are responses to their design. The second part topic is residential. I think you have designed many residential projects. In the same project, there is a variety of units such as Solidam and Mirador, from my observation, a variety of material and color are exist on the facade of the Solidam. It produces a kind of illusion that the characteristics of any materials are weakened, at the same time, there is mixed feeling of the whole produced—fragmentation of feeling. Could you illuminate the first design? What's the reason to design the different color and different materials on the same facade, what's the reason to response to the different unit inside. It's very interesting, I want to know.

NATHALIE: I think it's about trying to get away from any limit of modern architecture. Of course we already realize which is beautiful with the design of these building at the Delft. It was filling of any limit with a lot of big housing states. Yet we know we have to continue building large housing complexes, so full of different

示了内部的很多群体和房子。我们试图摆脱许多项目的限制，人们会说，这是我生活的地方，我不只是一个人在生活。虽然我可能真的一直生活在同一栋楼里，但我也可以体验生活在不同房间里的感受。

李：你所说的"不同的颜色是代表我住在这里，你住在那里"[*]是什么意思。客观地说，我相信我的感觉是我们能够承受巨大的体量对环境的压力，我认为它是分开的。体量是如何分离的，所以对环境没有太大的压力？我觉得这很好，我很喜欢。

娜莎莉：它实际上也不那么引人注目。

雅各布：我们回到它位于港口这个现实来去讨论，在那时，Silodam 是那块地最后被允许建造的住宅建筑。来源于工业建筑和海港活动的声音触碰着场地的边界。不被允许往海港外扩建，因此我们就让住宅中独立的个人因素和影响因素相结合，营造了一种活跃愉快的居住效果，也算是对周边场地、工业建筑以及集装箱的回应，这种多样的空间可以有多种的解读。

[*] 在该住宅项目中，建筑师用不同的颜色来暗示不同的户型空间。

section, but also showing that there are groups of people and groups of houses inside. We try to get away from any limit of a lot of projects, so that people can say again this is where I live and I am not just one person in sail. Although I may really live in the same building, I can also experience living in different rooms.

LI: What do you mean about different color have a logo about I live there, and you live there and mixture. Objectively I trust my feeling is we can be under the big pressures of the volume to the environment, I think it's separate. How the volume had been separated, so there is little pressure to other, to the environment? I think it's really good, I like it.

NATHALIE: It's also less obtrusive indeed.

JACOB: It has to do with the fact that it's in the harbor. And at that moment, silodam was like the last residential building that was allowed. It was the kind of sound, lies from the factory, and harbor activity, that was touching this site. You could not build outside, where the harbor begins, that's why we combine the individual element of housing with the kind of effect to make it the ecstatic housing in the city, and response to the industry, the harbor and stake of containers, you can read it in different ways.

李：是的，我们可以用不同的方式来解读它，会有不同的感觉，但它与其他住宅完全不同。我想是这样的。

娜莎莉：现在这种手法普及了，如今很多人都这么做了。这种做法变得越来越普遍。

雅各布：这也是一个非常特别的心理设计阶段。

李：在一次访问中，我看到了 WoZoCo 住宅和 Parlcrand。这两个项目中，都有很大的悬臂结构和桥梁结构。它是否被用于构建人与自然环境之间更紧密的关系。我认为它的内部结构特点并没有透过它的外观展示出来，它被木质和混凝土的表皮所覆盖。我想这个方法和荷兰馆正好相反，荷兰馆是开放的。这会让我产生一种错觉，会让我思考这是怎么做到的。这真的很让人惊讶，它非常强大。

雅各布：那是因为你没有看到。桁架使建筑形体更具有体验性，使人的体验感觉更加强大，因为他们只看到体积而没有看到结构。

李：我觉得这种方式可能很抽象。这不像现代主义建筑的一般原则：建筑遵循功能，遵循技术。

LI: Yeah, we can read it in different ways, different feelings, but it's quite different from the others. I think so.

NATHALIE: It's often repeated now, many people do this now. It's becoming more and more normal.

JACOB: It's very mental design period as well, it's quite special.

LI: In a visit, I have seen WoZoCo and Parlcrant. I think in the both project I observed that there are large cantilever structure and bridge structure. Whether it is being used for construction more close relationship between the people and nature environment. I think its interior structure specialties are not showed from its facade and it's covered by the wood skin and concrete skin. I think this method is just opposite to the Dutch pavilions. Dutch pavilions is just open, so I think to me there is an illusion, let methink about how to do it. It's really surprising, it's very strong.

JACOB: That is effect you do not see. The truss makes it more experiencing, stronger, because they only see the volume not the structure.

LI: I think this way maybe very abstract. It's not like the formal principle of modern architecture that building is following the function, following the technology.

雅各布：悬挑出来的结构主要是辅助结构，这并不意味着如果建筑被结构所支撑，它就是客观的，结构就是它所看到的那样。在这个案例里，结构支撑着这个想法，结构本身是不可见的，也没有妨碍这个想法。[*]

娜莎莉：我认为我们有两个方向，我们强调的是体积、空间。结构通常是多方面来考虑的，因为要结合所有问题，构成空间，而结构更是隐藏在墙壁中。当我们只有楼板的时候，结构就变成了房间的一部分，被特别设计的一部分，它们是和空间一块考虑来设计的，它们是空间节奏的一部分，也是空间体验中必不可少的元素。就像伊东丰雄的建筑，结构有时就是空间，他大概也是被柯布的多米诺结构原型影响的，只有楼板和柱子；也正如我们的VPRO办公楼，那是一栋结构作为主角的楼，立面几乎是看不见结构的。另一种是建筑盒子连接在一起，结构是服务于特殊目的的。

雅各布：我们都有都市化的成分，为什么形式是这样的，这是实现这个目标的客观的、非破坏性的主要方法。

李：可以这样说吗，这就像之前说的一样。技术和自然不能分开来讨论。也许技术和自然、房间是混合在一起的。在这个住宅里，不需要表达更强的结构。

[*] 建筑师在这里强调不能仅仅从某个单一层面来思考和判断结构的合理与否。WoZoCo 住宅悬挑结构的形成，是多个层面思考的结果，而非基于结构本身的必要与否。

JACOB: Structure is more supportive, this doesn't mean if architecture is supported by structure in, it's objective, you see, the structure is what you see. In this case, structure is supporting the idea, The structure itself is not visible, not disturb the idea.

NATHALIE: I think we have two directions, we really emphasized the volume, the spaces. Structure are usually hybrid to make them all, to make spaces, and the structure is more hidden in the walls also. And when we only have plates, that structure is really a room, part of the room, specially designed, it's a part of the rhythm, and the experience of the room. Like Toyo Ito's building, where the structure is space sometimes, which is influenced a lot by Le Corbusier, where there are just floors and columns. The way position the column and structure is also like our VPRO office building, where is really about structure of the building, the structure is almost invisible on the facades. The other one is sort of boxes connected together, that structure is more serving the special purpose.

JACOB: We all have urbanized component, why do form is like this, this is the main objective and distructive way to make it work.

LI: Could I say that it's just like the word before. Technology and nature can not be separated to discuss. Maybe technology and nature and the room are mixtures together. In this residential, it's not necessary to express stronger structure.

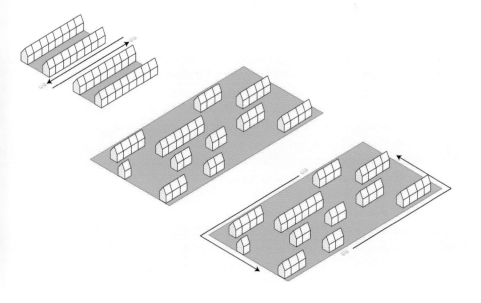

娜莎莉：我认为当结构变得更重要的时候，你也应该做一些在功能上更混合的项目，因为有一定程度的主次关系是很重要的，我们要意识到很多东西，墙、顶棚都可以改，唯一不变的只有楼板和柱子，而认识到这些会让整个建筑拥有更清晰准确的形状，这就是雅各布说的都市化的结合。事实上，就像市场大厅，它非常精确，很有趣，支持创作的投入。在一些住宅和办公室项目中，真的是关于创造大量的楼层、阳台和你可以进行设计的表皮，然后是如何优化结构以满足不同的使用。我认为它可能是经典的，但它是建立在如何构建的基础上的，建造的程序是什么，建造的重点是什么。

李：这有个关于哈根岛簇群住宅的小问题要问下你：为什么你要在屋顶和立面上使用相同的材料？通常情况下，屋顶和立面的材料是不同的，但是在你的项目里面都是一样的。同时，不同的房子使用了不同的材料和颜色。这与传统的方式截然不同，我想知道原因是什么呢？

雅各布：这是非常抽象化的开放的房子。所有房子都有屋顶风格，在这里一切都变得一样。我们想让它变成邻里街坊。房子的形状与细节无关。我们想把它变成一个村庄，并且与世界各地的村庄不一样。

娜莎莉：我们把它做成了很抽象的形状。这与技能无关，也不需要花很多钱在单一的形状上。

NATHALIE: I think when the structure gets more important you should do the projects that are a bit more hybrid in function, because you want certain majority. We have to realize that the walls, ceilings, anything can be changed, the only thing remain are the floors and columns, and with that the building could have more precise shape. It's more about what Jacob says it's a sort of a combination of urbanization. At facts, like the market hall, it's very precise, and such is fun, supporting the creation of the devoting. And in some housing and office projects, it's really about creating mass floors, terraces, surfaces that you can programme on, and then it's about how to optimize the structure to allow different uses. I think it's classical probably, but it's based on how to structure, what the programme in, where the emphasis.

LI: There is a small question just about Hagen island. Why do you use the same material on the roof and facade? Usually in tradition, materials on the roof and facade should be different, but in your project they're all the same. And the same time, different materials and colors are used in different houses. It is opposite way of tradition. What's your reason? I don't know.

JACOB: They are very abstract version open houses. All these houses have roof styles and everything becomes to the same. We want to make it become neighborhood. The shape of house is not about the detail. And we shift it to a village and the village is different from all around the world.

NATHALIE: We make it very abstract shape. It is not about the skill and the house is not worth money to spend on single shape.

雅各布：我们实际上是把钱花在了造园上。我们把它切成小块以得到更多立面，从邻里街坊和其他地区得到了更好的办法。

娜莎莉：这与风格和图像无关。

雅各布：屋顶的材料就是立面的材料。你可以把屋顶材料用在立面上，你也可以把立面材料用于屋顶上。

娜莎莉：它的所有材料最初都用于屋顶。

李：但是从立面上看，我认为所有的材料都是非常客观可行的。

雅各布：还有水槽，我们把水槽放在了底部。

李：我认为你们的概念是与众不同的。在你们的设计里，立面和屋顶是相联系的。

娜莎莉：我们也是年轻的建筑师和80后。每一个建筑师都是一样的。我喜欢玻璃、木材加砖，玻璃、木材加混凝土。我们开始在波兰项目中建造它们。我们的第一个建筑，我们最喜欢的房子有钢制玻璃和混凝土。

李：有一个问题，是关于鹿特丹市场大厅的。这里有一个关键词——内部化，也许设计市场的概念来自于这个词？

雅各布：不不，它是更有效的，你可以创造一个更大的拱形市场，整个物体将会变得更立体，房子的建造过程会变得更好，房子也会变得更好。

JACOB: We actually spend money on making a garden. We have more facades by chopping it up, and we get better ideas from neighborhood and the rest of district.
NATHALIE: It's not about style and illustration.
JACOB: And the roof material is facade material. You can put the roof material on the facade, or put the facade material on the roof.
NATHALIE: It's all the material that are originally used for usable roof.
LI: But from the facade I think it works in avery objective way.
JACOB: And the gutter, we put the gutter in the bottom.
LI: I think your conception is all different. In your design, facade and roof are connected.
NATHALIE: And also we are young architects and 80s. Every architects are the same. I like the glass, wood and bricks; glass, wood and concrete. We start to build them in the Poland project. Our first building, our favourite one is with steel glass and concrete.
LI: Last question, I think it's the topic about Rotterdam market. There is a key word-internalization. Maybe the conception of designing the market comes from the word.
JACOB: No no, it is more effective, you can create a bigger market with arch and the whole thing will become more steric. And if the construction of house is better, the house will be nicer.

房屋原本立面

当前场景

房屋原本立面

加大体量后的房屋模样

用玻璃重建房屋立面

体量扩张后的新立面

斜屋面上铺陶瓦并贴玻璃面砖

李：当进到你的中心空间时，我有一种非常强烈感觉。这种感觉类似教堂给我的感觉。一个教堂控制着所有的花朵、颜色和大小。它们关系非常紧密，这种感觉是非常开心和愉快的。我非常喜欢它。但是在第一眼看到它时，我很惊讶，我并不知道它是什么。

娜莎莉：是的，房子很漂亮。因为是商店的缘故，底部比较厚，但是总的来说，这几乎只是一样东西和一层厚度，一个房子。

李：就像你的屋顶和正面，我想问最后一个问题，非常简短。你能告诉我们创作水晶屋的最初理念吗？你为什么要这样做？

雅各布：我需要考虑那是在街边的公寓，同时那也是一个世界级品牌的旗舰店。我需要创造一些透明度。有一部分人喜欢老街区，另一部分人喜欢逛街购物。这两种想法创造了房子的解决方案。这是一个简单的疯狂想法，我们可以在技术和工程人员的帮助下创造它。我们做了很多立面调查。每一个人都喜欢玻璃砖。它是如此美丽。玻璃之美。

李：这非常好，我从你的身上学到了很多，非常棒。

娜莎莉：我们之后还可以继续联系，有什么问题可以相互咨询。

李：好的，非常感谢。

LI: In your central space when I come in, I have a very strong feeling. That's kind of feeling is how I feel about a church. Yeah, a church about controlling all the flowers, color and size. There is very close of relationship. The feeling is very happy and cheerful. I like it very much. But when I firstly look at it, I'm very surprised. I don't know what it is.

NATHALIE: Yes, it's very nice of the houses. It's a big thicker in the bottom because of the shops, but in general it's almost one thing and one thickness, one house.

LI: Just like your roofs and facades. I just want a last question, a very short one. Could you tell us about the original idea to create the crystal house? Why do you try to do it?

JACOB: I need to consider the accommodation of street, I need the commercial space to the world, I need transparency. There are some people like the old street and others like shopping. These two ideas create a solution of the house. It is simple crazy idea and we are able to create it with the help of technology and engineering people. And we do a lot of research of facade. Everybody loves the glass brick. It is such a beautiful thing. The beauty of glass.

LI: It is very good and I learn a lot from you. It is extraordinary.

NATHALIE: We can also send some questions to each other.

LI: OK, thank you very much.

对谈法兰馨·胡本（Francine Houben）

访谈时间：2016 年 12 月 09 日

法兰馨·胡本毕业于荷兰代尔夫特理工大学，目前是 Mecanoo 建筑事务所的创始人和创意总监。胡本的作品类型非常广泛，包括大学、图书馆、剧院、住宅区、博物馆和酒店等项目，曾于 2014 年获得建筑师学院年度女子建筑师的奖项，于 2015 年获得 Prins Bernhard Cultuurfonds 奖等。

李：第一个问题，请先让我表达对（代尔夫特理工大学）图书馆和大礼堂的感觉。这两座建筑震撼了我，它们各自有着如此鲜明的特色，却能如此协调一致。众所周知，大礼堂在1966年就建成了，三十年后图书馆也建成了。它们属于不同的年代，也基于不同的设计理念。显然，大礼堂的设计原理是形式跟随功能和技术，它的风格是粗野主义。而在图书馆的设计当中，我可以找到对人与环境之间关系的尊重。在你的书（《Mecanoo Architecten: People Place Purpose》）中有两段话描述你的设计思想。能谈谈你的创意来源是什么吗？是否基于对环境的深层次理解和对人类感知的判断？

胡本：你对这座建筑的描述非常好，我认为你去看图书馆的设计是很重要的。作为这个建筑的设计师，我认为把这两个建筑联系在一起是很必要的，你必须去做这件事。巴克马（Bakema）在1960年代设计了大礼堂，它是完全不一样的，体现的是粗野主义的未来，它和混凝土的发明与建筑的结构有关。当然，1970年代我在这所大学里面学习，我非常喜欢在这所大学里面学习，但是没有在这里，我也在建筑系的楼里学习过，它也是由巴克马设计的。老实说，我非常喜欢它，但是那里没有室外的空间，没有校园的感觉，建筑外也没有可以坐的地

LI: As for the first question please let me express my feeling of the library. My first impression is very shocking about the two buildings with such distinctive characteristics could be so coordinate together. And obviously Aula was completed in 1966, 30 years later, library was built. I think they belong to different times, based on different design conception. Obviously the design principle of Aula is the form following the function and technology, and its style is brutalism. To the opposite, in the library design, I could find the respect to relationship between people and environment. There are two paragraphs in your book to describe your design idea. What is your creative source? Whether it is based on the deep understand of environment and the judgment about human perception and feeling? Could you talk about it?

HOUBEN: It's nice that you describe the building, I think it is important if you look about the design of the library. As the architect of the buildings, I think they are father and daughter and it was important to put the two buildings together, you also has to do something. Bakema designed the repertory Aula in the sixty's, it is a totally different side guy, it's the future about the gross, about also the invention of concrete and the structure of building. And of course I have studied in that university set in the 70th, and I'm very much like to study in the university in Delft, but not in there and I'm also studied in faculty architecture which is also designed by Bakema. To

方。当我们开始设计图书馆的时候，我认为非常重要的一点是，巴克马的建筑就像是在月亮上的一艘船。因此你不能单纯地把另一栋建筑放置在这个建筑旁边，这会影响大礼堂的主导地位。我们选择将建筑设置在一个大草坪下，然后将草坪升高，这样就建造出了图书馆。这样的做法也形成了一个我一直渴望拥有的室外空间，你可以坐在这个室外晒太阳，就像在山上一样。假如我做了一个曲折的屋顶绿化，我想这样只能让小狗上来，但是人走不上来。我认为不像巴克马的设计也会更好。这就是我为什么说"他们是父女"的原因了，他们现在是一家人了。

李：是的，我明白。第二个问题是关于屋顶草坪的边缘升高，考虑到周边环境的影响，是否对传统的原则有所解构、联系或者反映，深层次的构建传统原则是跟随在技术和功能之后的。我认为它打破了常规的原则，常规的原则是追随技术和功能。

胡本：如果你在同一时期比较这两个建筑，我是同意的。但是大礼堂也被很快遗忘了，我认为这也不是形式仅仅追随功能。但是另一方面，我认为我们所做的图书馆也应该提供给人们一个休憩娱乐的空间。作为一个建筑师，我们在学

be honest, I enjoy that very much, but there was no outdoor space, was no campus feeling, was no place to sit outside of building. So when we start to design the library, I think it is extremely important that I felt the building of Bakema was like the ship from the moon, So next to the building you cannot just put a building, it's such dominant strong ego, so that's what we said we put and made also Bakema by putting it in the grass, then elevate the grass and make a building and become the library. It is also extremely my longing for having outdoor space where you can sit outside in the sun, happens more in the mountain. Then if I make a flight green grass area, I'll say it's just for the dogs, but it's not inviting. So I think without the design of Bakema the building would become much better. So that's why I said they are father and daughter, they are family now.

LI: Yes, I understand. And the second question is raising the lawn on one edge, expecting the environment reason, whether there is a deconstruction or relation or reflected to the traditional principle, the deep construction to the traditional principle is form following the technology and functions. I think it broken the formal principle, the formal principle is follow the technology and functions.

HOUBEN: If you compare two buildings at the same time, I agreed. But also the Aula was forgotten quickly, it's not form just follows function I think. But in a way, I think the library we designed should provide the place where people play. As architects, we learnt purpose driven methods in school and read so many great

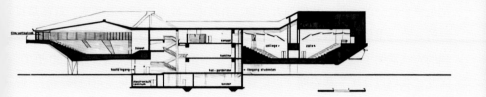

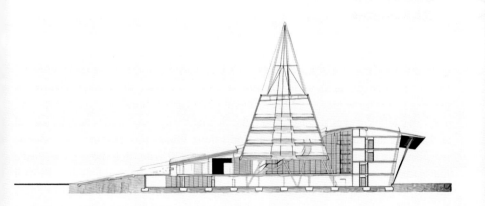

校里学到了形式追随功能的方法，我们看过很多著名的建筑设计案例，还有很多充满了灵活性、弹性的案例。但是我从职业生涯的 30 年里学习到的是，目标总是在变化的，在现在这个时代，世界的变化是很迅速的，技术也在不断改变。这就是我把功能放在第三位的原因，这对我来说是很重要的。在代尔夫特、深圳、高雄或是波士顿，设计的方法是完全不一样的，因此，地域是我们需要考虑的第二重要的因素。第一重要的因素是人。你和我或许会有点不同，但我们都是整个社会的一员，我们拥有感知，有耳朵，能感知到声音和光。这就是我说的，第一重要的是人，第二是地域，第三是功能的原因。我认为图书馆以极端的方式证明了这一点，因为当我们设计图书馆的时候，距离现在已经 20 年了。因此，我们设计了一些空间，以应对将来的改变，这些变化是可以预测的。世界会变得更加的数字化，书的目录会消失，书将会变得更少，而数字化的信息将越来越多，特别是在代尔夫特理工大学。还有一些我所不能预测到的改变，因此我们创造了可以迎合这些改变的空间，而这种变化可以顺利地发生，因为我们更新了图书馆的内部。

李：对，正如我们所知，我认为你所说的对我们来说非常重要，人的感觉或者感知是不稳定的，技术和文化在不断变化。

works, some flexibility. But what I learn by working now more than 30 years, the purpose always change because you know especially in this time the world is changing, the technology is changing. So it's why I put purpose on the third place, it is extremely important place for me. It's totally different that I can make in Delft, in Shenzhen or in Gaoxiong, or in Boston, so it's the place the second. The first is people. You and me maybe a little different but we are all kinds of society, we have sense, we all have ears, acoustic and light. So that's why I say, first is people, second is the place and third is the purpose. And I think library proves that in an extreme way, because when we did design the library, that's 20 years from now. So we made the space that was prepared for changes and some changes he could predict. He said the world could be more digital, so the catalog will go away, so there will be less book and more digital, especially in Delft University. And the part of the change I do not know so we created the space that could ingratiate the change, and that happened because we did update the library interior.

LI: Yes, and as we know, I really think what you say is really important to us, and the feeling or perception of people is stable, technology and culture are changing.

胡本： 是的，但是有些与人密切相连的文化特性是一样的。或许你有着中国的根源，我有着欧洲的根源，但同样的是，我们都喜欢照顾孩子，无论贫困或富裕，人们都有着一些相同的价值观，这些价值观没有政治上的联系。你知道的，无论你有什么样的政治背景，想去照顾自己孩子的心是不变的。我们想去照顾好我们的父母，想变得健康，还有一些其他的因素，就像快乐，这些都与此价值观相关。所以，我总是说朝着最基本的价值观前进，因为我认为这个基本的价值观就是建筑设计思想的基础，而且非常重要。建筑设计应该触及人的所有感官，并且在这方面多下功夫，因为我们都有同样的感知。

李： 是的，我也这么认为。据我所知，你和西扎有过合作经历，在过去的几年里，我在西班牙的里斯本和波尔图看到过他的很多作品，他是否对你的设计思维有很大的影响？

胡本： 对我来说，他非常重要。因为在 80 年代，我们一起创立了工作室，我们那时候很年轻，大概 27 或 28 岁。他在工作室工作了两到三年，但我从他身上学到了一种强烈的目标驱动性和逻辑性思维。但是正如你所知道的，他设计过两座房子，在 Egg 的公园里两座非常特殊的房子。那时候他在我们的工作室，我们是他的法定建筑师。但是例如你要在荷兰做建筑，这里的规范特别具体，你已经被规定了：起居室应该是 24 平方米，厨房是 11 平方米，等等。但是他所设计的客厅有点小，虽然我认为在代尔夫特理工大学里不存在应该修正的僵

HOUBEN: Yes, but something of culture, which are relate to people, are even the same. Maybe you have Chinese origin and I have European origin, but we both like to take care of children, even some are poor some are rich but they all have the same values, and those values they have are not political connected. You know no matter what kind of political background you have, you all want to take care of you children. We want to take care of our parents, we want to be healthy, also many different things like happy, it's very much about this basic of values. So I always say forwards to basic because I think this basic values are forming architecture and also extremely important. So it's the factor that architecture should touch all the senses, and have much to do because we all have the same senses.

LI: Yes, I think so. And as I know, you have working experience with Siza, I have seen a lot of his work in the past years, in Spain, Lisbon and Porto, is there a great influence to your design thinking?

HOUBEN: For me, he was important. Because it was in the 80's, we were, you know, we have already made up the office, I was still very young, I was 27 or 28 years old whatever. He works for 2 or 3 years in our office working on the project, but I learn from him very much purpose driven and a very much logical thinking which I was told in the faculty of architecture in Delft. But I remember you know he

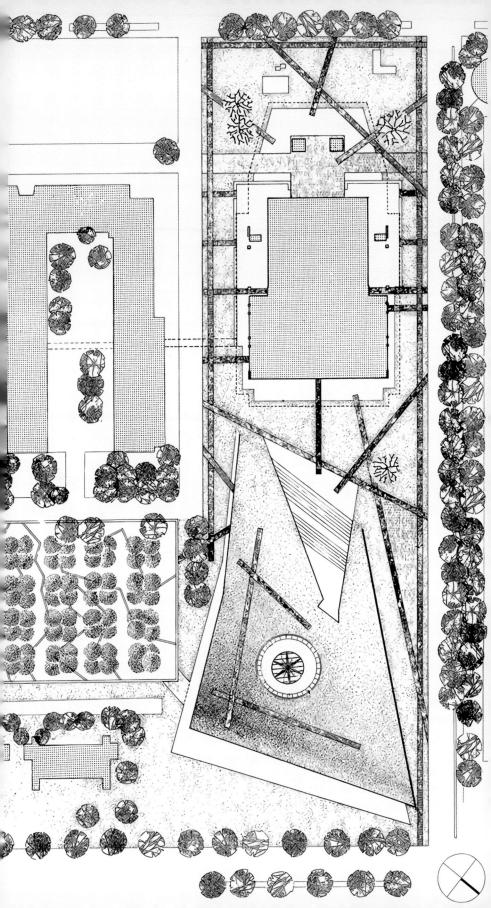

化性因素，但西扎只是将外形做的稍微大了一点，这是源头[*]，使你的头脑中的形式和想法获得了自由。对我来说，我在他身上学到的是超越形式的，更加自由地去思考。

李：我从彼得·卒姆托那里听到过同样的一句话，他不在乎形式，只关心感知，关心人与人之间的关系、建筑的外观、距离以及材料，我认为这非常重要。但是在中国，一些建筑师真正关心的是建筑形式，他们只关心建筑是否美观，但我认为这不是最终的目的，这是错误的。

胡本：我反对相片类的建筑，因为单纯的相片是不能带给人感知的，光线是怎样的，声音是怎样的，材料带给人的真实的感觉，地板是怎样的。

李：正如你在书中所说的，图书馆里面的巨大锥体不仅仅是建筑的结果，也是一种符号，所以我想问一个关于符号的问题。建筑师往往在设计过程中进行逻辑意义的建构，并将形式和空间背后的意义转化，建筑师如何才能清晰地将意义转化为符号？我们是否应该用建筑语言来表达它，而不是使用其他形式的艺术语言？

* 西扎的意思是，摆脱空间被面积严格控制后，使设计变得更具灵活性。这是设计具有灵活性的源头。

sitting there, for instance he has made two houses, very special houses in the park in the Egg. He was working in our office, we were his legal architects. But for instance like if you make in Holland, maybe it is too specific but you have regulation and the living room should be 24 square meters, and the kitchen is 11 and so on. But he made the design and the living room was a bit small, in my tradition of Delft University of Technology that I have to make change, but he made just a little bigger, and it was for source, so give freedom to form and in your mind. For me, what I learn from him was that it is very important to think in a more free way of form.

LI: I think the same word I have heard from Peter Zumthor, he doesn't care about form, he just cares the feeling, the relationship between people and facade, the distance and the material, I think it is very important. But in China, some architects really care the form, they just care whether the buildings have good appearances, but I think it is not final purpose, it's wrong.

HOUBEN: I against photoshop architecture, because photoshop makes no sense, how the light is, how the acoustic is, what the material reality is, what's on the floor.

LI: And as you say in the book, the large cone in the library is not only the result of the construction, but also a kind of symbol, so I want to ask a question of the symbol. Architects often carry out the construction of the logic meaning in the process of design, and transform the meaning behind the form and space, how could the architect transform the meaning through the symbol clearly? Should we express it in a construction language, rather than using other forms of art language.

胡本：图书馆有了这个美好的形式，它给人感觉像是一处风景，或者是让人能交流的一个地方。其实我非常担心这个坡屋顶过于陡峭，但是它仍旧让人觉得像是一道风景。但在这之后，这种景观的模式成为代尔夫特的符号。它就是代尔夫特理工大学的代表，我想让他们之间产生一种联系，感性的形式加上理性的形式。于是锥这个形体就出现了。它同时也产生了一种自由感，因为我们和结构工程师一起设计了这个锥体。这是个自由的空间，这就是我在书中所提到的形体的复杂性，但是它的普遍存在形式是自然形态和结构形式的组合。就像我们在高雄的项目，就像这种形式，但是这种地景建筑非常具有标志性。如果你看我们在韩国设计的建筑，我们是水平向的在做建筑；同样地，在荷兰的博物馆也是一种组合，那是一个倾斜的风景，是一片墙，一块大石头更突出了其形式。

李：当我进入图书馆内部的时候，由入口对面一侧的大书架营造出来的文化氛围给了我很强烈的印象，它是整个空间的基调。这给了我一种非常强烈的感觉。同时，锥体上面的光影变化，使得大厅充满了活跃的氛围。这让我想起了由彼得·卒姆托设计的瓦尔斯温泉浴场。

胡本：我没有去过那里，我只从照片中了解过浴场。我总是跟我丈夫说我有一天会去那里的。

HOUBEN: With the form of what was nice about the library itself, it is really like a landscape or free form with a landscape for communication. I'm very worried about it should not be too steep but it still feels like a landscape. But then, what betrays that value, the landscape is a symbol of technological beauty, and it is Delft university, so I want to prove my combination, the emotional almost form with the rational form. That is the cone, it was also given the freedom, because we did it also together with the structure engineer. The space is free, this is the complexity of form that I mentioned in the book, but its form that is often in the world is the combination of the natural form and struct form. Like the building in Gaoxiong, it was like this form, but the landscape is very flag. If you look at our building in Korea, we made the building like this, the horizontal. The Dutch museum over there is also a combination, that's a sloping landscape, a wall, and a big stone gain the form more.

LI: And when I get into the interior, the strong impression is the culture atmosphere, created by the large book shelf, at the opposite side of entrance, which is the basic tone of entire space. I think it gives me a very strong feeling. At the same time, the changing light and shade on the cone make the hall full of energetic atmosphere. It reminds me the Vals Bathroom, designed by Peter Zumther.

HOUBEN: I've never been there, I only know from pictures. I told my husband I always want to go there one day.

李：那是一个很棒的地方。那里给人的感觉非常好，光线从顶棚上照入室内。

胡本：有两样事物可以做个对比，我总是把建筑看作一部电影，把建筑看作一组序列，然后你用你的身体穿过这组序列。你首先走上台阶，然后把门打开，你进去之后看到最里面是蓝色的墙，在右手边是桌子，它是一系列的空间。我感觉建筑内的灯光是那么的美丽。这也就是为什么我决定设计一个完整的玻璃幕墙，因为光线是美丽的。我将多功能的办公室放到前面，然后设计了一面不同的玻璃。在西侧，它面对着巴克马设计的礼堂，它们之间距离很近。人们可以坐在那里晒太阳，就是这样。

李：从你的话中得知，所有的元素不仅仅表现在工程的意义上，还表达在人的感觉上。显然，图书馆也是个成功的绿色建筑，绿色屋顶在控制大厅温度上有重要的贡献。你可以解释一下气候墙的技术吗？它是双层表皮吗？

胡本：你所说的在立面上的窗户，当然，建造这个是非常重要的，对于被日晒的屋顶绿化来说也是非常重要的。因为它是真正的玻璃，这使建筑能够更好地适应气候的变化。而立面是由三种玻璃组成的，我知道的不是很确切，这三种玻璃分别应对三种气候。

李：那么关于冷储藏和热保温呢？

LI: It also is a good one. The feeling is really good, light comes down from skylight.

HOUBEN: There are two things to compare, I always look into buildings almost like a movie, a kind of sequence, and then move your body through buildings. So at first you make the steps, and the doors open, then you go there, what you see at the end drab is the blue wall, on the right hand side it's the desk, it's a sequence of spaces. I totally feel the light is so beautiful in the building, that's why I begin to build the total glass. I put the flexible offices in front and design a different kind of glass. At the western side, it faces the Bakema's auditory, it's close, people can sit under the sun, that's all happening.

LI: Get a conclusion from your words, the all elements are used not only in the engineer meanings but also in the feeling perception meanings. And obviously, the library is also a successful green building, the green roof have a significant contribution to control the hall's temperature. Could you explain the technology of climate wall? Is it a double skin?

HOUBEN: The windows you mean, on the facade, of course it is very important by making this, also the landscape that insolated is really like this. Because it is real glass, that makes the building very stable in climate. And the facade is made up by three kinds of glass system, I don't know it exactly, but they face three kinds of climates.

LI: And how about the cold and heat storage?

胡本： 冷储藏和热保温是首位的，我认为它非常重要。在屋顶上我们没有设置保温材料，我们把它做成比较小的设备然后部分藏起来。如果不这么做，会有一个很不美观空调设备暴露在屋顶绿化上，因此这么做是非常必要的。

李： 是的，我明白你的意思。根据我的观察，空调机是否隐藏在柱子的表面之下？我看到中间有一个柱子。

胡本： 柱子是非常重要的。这是屋顶，这是地面，这是柱子，这样，如果你看着这一部分，它就像这个。这是空调，这仅仅是起着承载的作用。这里我们把光引进来，所以这里比较宽，这就像一阵风，它带着屋顶。这里像这样，这里的部分是这个，所以空气可以进入。它们一起产生作用，这里也好像正在发生变化。我说柱子起着承载的作用。带来了光线，然后这是顶棚。（作者注：她好像在纸上画一些东西去解释。）

李： 是的。第二个项目是位于传统街区的代尔夫特火车站，它周围都是旧建筑。我认为在你的设计中，有一个城市设计得概念和分析方法，来处理新车站和旧环境之间的关系，使它变成了周边环境的有机组成部分。能谈谈你是如何考虑这个建筑的吗？这个新车站是一个非常大的车站，它的四周围绕着规模很小的房子。

HOUBEN: The cold and heat storage is in ground, I think it is very important, we don't have insulation on the top of the roof, so we made it in little parts to be hidden. Otherwise there will be a landscape spy and an air condition coming out which is ugly, so hiding them is very important.

LI: Yes, I didn't see any machine outside. I know what you mean. According to my observation, is the air condition hidden behind the surface of column? I see there is a column on the middle of hall.

HOUBEN: The column is very special. This is the roof, this is the floor, this is the column, if you look the section this way, it's like this. So here is air condition which is just carrying, here we put light up, so this is fat, this is like a wind, it's carrying the roof, and here is like this, the section of here is this, so the air could go in. It's column to do and here is like to coming up. I say the column is carrying and giving light, and this is the ceiling.

LI: Yes. And the second project is about the Delft train station, which is located in the traditional district, surrounded by old buildings. I think there is an urban design conception and analysis method in your design, to deal with the relationship between new station and old environment and to make it become an organic part of environment. Could your talk about how to deal with it? The new station is a very big one, and the surrounding are very small houses.

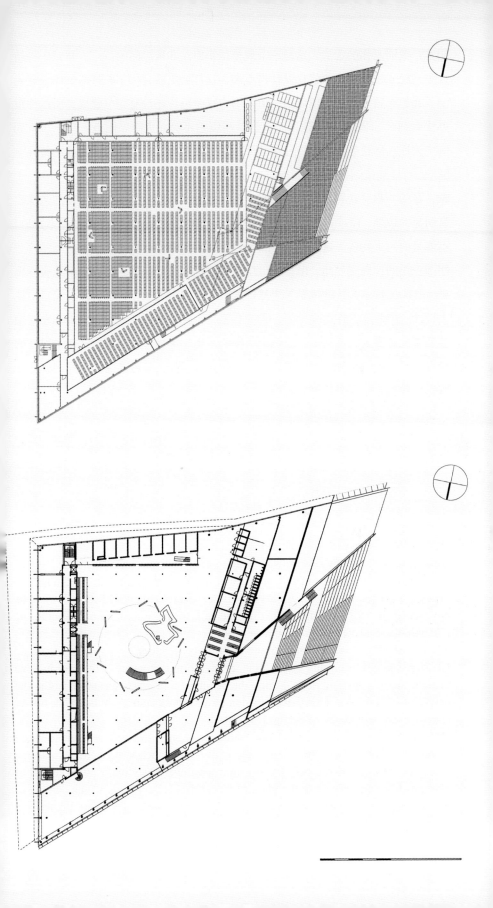

胡本：我同意你的看法，这是一个规模很小但是非常漂亮的城市。我试图将新火车站融入这个城市，将它设计得不高，这样会更适合代尔夫特的整体景观氛围。我们在这方面做得很成功，但是因为这个设计还没有完全建成，所以你没办法看到。[*]同时创建了这两个大厅，所以它不会在火车站附近。接下来看到的是大厅。部长办公室设置在市政厅，那里创造出一种几乎持续开放的公共空间，它可以连接到城市的各个地方。我对火车站感到无比自豪。

李：是的，我喜欢它。从我的观察来看，首先，从屋顶的中心到屋顶的边缘，有明显的坡度。屋顶边缘的高度几乎跟旧建筑的高度持平。（胡本：我不知道，可能是的。）我认为这或许可以减轻对周围的压力。

胡本：是的，确实如此。但是我也想把我对代尔夫特的所见所想自由地表达出来。这（《代尔夫特的风景》）是维米尔的一幅著名画作，它启发了我自由地去设计。因为如果你这样做了，我的意思是，在某种程度上，每个人都喜欢用玻璃来建造建筑的外立面，玻璃到顶棚的距离非常高，这不能给你一个很好的视野，所以这就是我们试图创造的，空间开始变高，又再次变低，我们创造了城市的景观。

* 访谈时代尔夫特火车站正在进行二期扩建。设计将车站与新的市政厅结合在同一体量中。

HOUBEN: I agree with you, it's a small scale but very beautiful city. We designed it not high to integrate the new railway station into the city, because otherwise it didn't fit in the urban landscape of Delft, and I think we succeed in that, but you can not see because it is not all finish yet. At the same time, we create this two halls, so it will be not around the railway station. Next it's the hall, for the minister's office set in the city hall, there creates a kind of almost continues open public space, connecting it to all sides of cities. I am extremely proud of the railway station.

LI: Yes, I like it. From my observation, firstly, from the center of the roof to edge of roof, there is obvious slope. I think the height of roof edge is nearly equal to the height of old building. (HOUBEN: I don't know, but maybe.) And I think it may reduce pressure to the surroundings.

HOUBEN: Yes, that's true. But I also wanted to express freely what I had seen and thought of Delft. It was a famous painting of Vermeer which It inspired me to design freely, because if you did that, I know, in a way, everybody made buildings all of glass, and the distance from the glass to the ceiling is very high, which can't give you a nice view. So that's what we try to create, and then the space goes higher, and goes lower again, we create the view of the city.

李：所以我认为屋顶边缘的高度是将建筑融合进周边环境的重要元素。这让我想起了在巴黎由鲍赞巴克设计的住宅。（**胡本**：一个住宅建筑？）那是一个大体量建筑，他把整个体积分成几个垂直的小单元，我认为这种做法也减少了建筑对人们和周边环境的压力。从这两栋建筑里，我得出了同样的结论。最后一个是高雄卫武营国际艺术中心。我还没去过那里，但是从你书里面的材料和照片看出，（**胡本**：那里景色很美。）它位于在一个公园的中心，有一个从主体分离到边缘的斜坡，与树木的轮廓相连，从很远的距离我们就可以看到非常舒服的艺术地带。你能谈谈这个吗？

胡本：好的，我认为这个想法是很重要的。高雄的文化[*]，我喜欢的正是构造上的，它非常的暗，当然，中国文化是喜欢在晚上采用人工照明，人们喜欢这样。但是这里经常发生地震，经常有台风，还有暴雨，等等。当然还有一部分军事基地，这样做就可以把所有的兵营都隔离开来。我记得第一次来的时候，那里有兵营，人们把车停在室内。那里还有很多榕树，我的设计灵感大部分都受到榕树的影响。从树那里你能感受到生长的力量，榕树的根很多很长，从树上一直垂到地下，这是一种特殊的形式，一种延续，是只有中国才有的形式，人们

---

* 高雄地处亚热带地区，室外、半室外的活动场所——例如榕树下的乘凉空间——成为人们生活中很重要的组成。

LI: So I think the height of the roof edge is important element to the surroundings. And this reminds me a residential in Pairs designed by Portzamparc, (**HOUBEN:** Yes, a residential building?) it is a big building, he separates the whole volume into several vertical slim units, I think it also reduces the pressure to people and surroundings. From the two buildings, I have a same conclusion. The last one is the Weiwuying Art Center. I have not been there, but from the materials and pictures in your book, (**HOUBEN:** Its view is so beautiful.) it's in the centre of the park, there is a slope from the central roof to the edge and connected to the outline of trees, and from long distance we can see the up and down outline. Could you talk about this?

HOUBEN: Well I think this idea is the most important. The culture of Gaoxiong, what I like is very tropical, it's very early dark, of course the Chinese culture likes artificial light in the evening, they like it. But it's also a lot of earthquake, typhoon, rain and so on. And of course there is a formal military compound, so this will isolate all the barracks. I remember coming there the very first time, there was barracks and people just park indoors. And there are lots of banyan trees, so the design is very much inspired from the banyan trees. From the trees you can learn the growing power, the roots of banyan trees are many and long, and they hang from the tree to the ground. This is a special form, a continuation, which is only a form in China. People like to perform under banyan trees. This is an informal performance. So the

喜欢在榕树下表演，这是非正式的表演。你看到的我的建筑设计理念就是这种形式，我将这种表演形式反映在建筑当中。当然，我们建造了人们要求我们所设计的歌剧院、音乐厅以及剧场，这些都是正式的艺术表演形式。我也非常喜欢非正式的艺术表演形式，因此我们想为那些没有票，不能进入音乐厅参观的人们创造空间。如果回顾中国戏剧的历史，它就是在大街上表演的。所以我们怎样才能把公共空间和大体量建筑联系起来呢？我带着这个思考去设计建筑。我喜欢去高雄，我非常喜欢港口城市，我总是和很多移民在一起，（**李：**那是一座很悠闲的城市。）因为那里的港口，就像我一直住在那里一样。我们所做的，我认为了解艺术表演中心里面的榕树是很重要的。这个建筑体量很大，我们把屋顶盖在建筑上面，就像榕树一样，屋顶连接着公园的那一部分就变成了露天剧场。很重要的是，这让你更加充满激情，因为在这个空间你可以看到戏剧。这就是我的设计理念，这也是我想创造的。在晚上，这个空间是白色的，我们在变化的光影之下活动，也是在中国文化的氛围当中活动，你可以调整光线，所以我们可以达到我们想要的效果，这种状态是由当地街道上的主要建筑所塑造的。榕树地块，它是当地榕树的一部分，在街道的尽头是当地的主要建筑。那棵我们想坐在它上面的整个大榕树，它就像一条船。

idea you see is also this form, I reflect this form of performance in architecture. And of course we build what people ask us to design, the opera house, the concert hall, recital hall, and the playhouse, that's all formal form of arts. And I also very like the informal form of arts, so we want to create space also for people who don't have tickets and are not allowed to visit the concert hall. So if you look at the history of Chinese opera, it's from the street, so how can we connect the public space with the big building? I design architecture with this question. I love to go to Gaoxiong and I very like the harbor city, I always with more lots of immigrants, (**LI:** It is a relaxing city.) because of the harbor, it's like I've always lived there. So what we did, I think it is important to understand the banyan tree in the big perform art centre. And you know it is big, it's a big building, and we put the roof over it, like the banyan tree, and where the roof touches the park, it becomes the open theater. It is important that you are really tropical, because in this space you can see theater. And this is the idea of my design and what I want to create. In the evening it makes the building totally white, we are playing with the changing of light and also playing in the culture of Chinese people, you can change just the light, so we can adjust what we want, and it is made by the local chief building in the street. The banyan plats, it's part of the local banyan tree, and at the end of the street are the local chief buildings. So the whole banyan plats where we want to sit on, is built like a ship.

李：我认为这艘船和榕树是建筑的两个主要元素。

胡本：是的，我之前在那里。大概是在三四个星期之前的一个夜晚，我想知道这东西是否还活着，我和十五个人站在一起，因为那是声学景观。[*]在晚上，那里充满了爬行动物，听起来很美。想象一下，在未来，老人们被榕树包围，这现象就在这里发生，这是很正常的。这个建筑差不多建成了，但是我想它明年才会正式完工。我认为演出将会在 2018 年开幕。

李：我认为这是一个非常复杂的项目。

胡本：是的，这很难，但也没那么难，正如我们在同一时间做着另外两个项目一样。但如果你说我们的建筑很复杂，实际上并非如此。人们都做得很好，屋顶做得很好，榕树广场做得很好，剧场内部空间的形式很漂亮。

李：这幢建筑与周围的环境营造出了很好的关系，我喜欢它。还有个问题是，你能评价一下关于代尔夫特理工大学建筑教育体系的吗？

* 建筑师在描述建筑存在的状态，是否真正能成为当地居民生活的一部分。

LI: I think the ship and banyan tree are the two main elements of the building.

HOUBEN: Yes, I was there. I think three or four weeks ago, in the evening, I wonder if the thing is alive, and I stand with fifteen people, because it is acoustic landscape. In the evening, there are all creepers, it sounds so beautiful. And imagine in the future, the old people are surrounding by the banyan trees, it happened here, it's normal. It almost finished but I think it will be finished next year. I think the grand will open in 2018.

LI: I think it is a very complex, difficult project.

HOUBEN: Yes, it is difficult and also not difficult, because for instance we do the project the same time with another project. But if you say our building is complex, actually it is not complex. People do very well, the roof is well made, the banyan plaza is well made, the form of auditory is beautiful.

LI: The building makes a good relationship with the surroundings, I like it. And the last question is, could you have an evaluate of the Delft architecture education system?

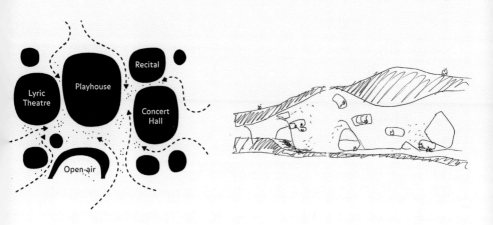

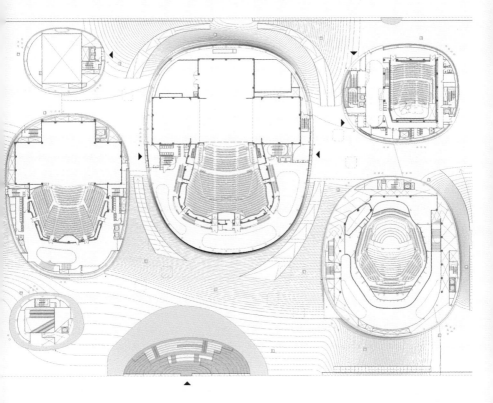

胡本：我最近没有在代尔夫特进行演讲或者授课，并不知道那里具体的教育体系。我只谈关于代尔夫特理工大学，但是我认为这些事情在任何大学都是非常重要的。我们不应该单单通过照片或者屏幕来看这个世界，在屏幕的背后有着太多东西。我认为学生们在走在街上的时候应该学习，体验身边建筑带给人的感受，还有建筑使用者的感受，所以我希望学生们更能感知身边的事物，而不是通过电脑屏幕观察这个世界。他们应该在电脑屏幕和真实生活之间找到平衡点。年轻的一代总是美好的，因为他们代表着未来。这是关于物质性的，你谈到了彼得·卒姆托，他是在乎物质性的，在乎细节的，你知道什么在这，这是什么，那是什么。

李：最后一个问题。根据你在荷兰的背景，当你修复非常复杂的建筑时，你最看重的是哪一方面，是直觉还是事物之间的联系？

胡本：我认为分析和想象两者之间的结合是很重要的。正确的选择和感情，但也应该意识到想象力是非常基础的，这是根据我以往的经验总结出来的。

HOUBEN: I have not been recently giving lectures or giving teaching in Delft, so I don't know exactly their system. But I think it is extremely important of any universities so I'm talking just about Delft. That we should not photocopying the world, there is too much behind the screen. I think the student should learn when they walk in the street, experience many buildings, experience the feelings of the buildings as well as the feelings of building users, so I want the students to feel things around them more than see the world through a computer screen, they should find the right balance between computer screen and real life. The young generation is always great, because they are the future but they should not photoshop the world. It's about the materiality, you talk about Peter Zumthor, it's about materiality, it's about details, and you know what is here and what is this and what is that.

LI: The last question. According to your background in Holland, when you fix the very complex building, which one do you relay on the most, the intuition or the relation?

HOUBEN: I think the combination of the analysis and the vision is important. The right choose and the emotion, but also be aware of that vision is very basic, it's almost experience of what I have done before.

对谈卡斯·欧斯特豪（Kas Oosterhuis）

访谈时间：2016 年 12 月 23 日

卡斯·欧斯特豪是荷兰代尔夫特
理工大学建筑与建成环境学院教授、
博士研究生导师，同时是 Hyperbody
and the Protospace 设计和工程协同研究
所的主任，荷兰 ONL 建筑师事务所负
责人。他的研究方向是交互式的建筑、
建筑物与环境的实时行为、居住建筑
概念、协同设计和参数化设计。

李：在国内的时候，我接受的建筑教育理念也受到了现代主义建筑的影响。现代主义是一个基本的原则，但它好像停滞不前，因此当我看到国外建筑的形式时，我觉得它很特别，很好，但是我怎么能理解它呢？我不知道为什么要设计这样一个建筑，也不知道设计师设计它的方式。另一种情况是，有些人想要打破这些原则，但仅仅是打破原则却不能建立新的秩序。在我读了你的书（《Towards a New kind of Building》）之后，我找到了一些答案。现在它仍然是一个混乱的状态，如果建筑的形式和空间是新的、特别的，人们就认为它很好，特别是如果它来自一个外国设计公司。人们盲目地相信它应该被追随，并将成为时尚，但我认为这只是一个表象。

卡斯：从国外哪里？

李：从欧洲和日本，人们可以欣赏美丽的外观，但他们不明白为什么这样设计或设计背后遵循怎样的思维过程。

卡斯：嗯，我个人认为它来自于内部，我不认为它必须来自国外。这更像是一种心态，在你描述的这段时期之后，从40年代到70年代，在荷兰有一个逆转，更多的是传统手工艺。你仍然会看到一种倾向，人们喜欢看更多的工艺品。现在，我们的方法是，让我们进行定制，使用机器人技术、数字设计——这就是新工艺。它是一种工艺，因为它直接关系到材料，你可以看到设计、生产，它与材

LI: When I was in China, the educate principle which I had been educated is also affected from the opinion of modern architecture. I think it's a basic principle, but it hasn't developed, so when I see the form of the building abroad, I think it is special and good, but how can I get it? I don't know the reason to design a building or the way that the designer to produce it. On the other hand, another situation is that some people want to break these principles, but it is just broken, and they didn't know how to establish new order, so it caused a confusing situation. They always consider good if form and space is new or special, if it is so, it must be good and becoming the trend. But in my opinion, it is just a seeming. After reading your book *Towards a New kind of Building*, I seem to find some answer.

KAS: Abroad from?

LI: From Europe and Japan, people can appreciate the appearance and beauty but they don't understand why or what thought process is followed behind the design.

KAS: Well, personally I think it came from within as well, I don't think it has to come from abroad. It's more of a mentality, I think, so that after this period that you're describing, from 40s to 70s, there was actually a reverse in Holland, going more to traditional crafts. You still see a tendency, people like to see more crafts. Now, our approach is, let's go to customization, and using robotics, digital design -

料直接相关。你不必把目光从中国移到国外，我的意思是，你可以从内部做一些事情。那不是你可以从国外引进的东西，而是你必须要内化的东西，你必须有同理心，有数字设计方法。这不是从国外引进的问题，而是内部的问题。

李：是的，你说的很有道理，我此行的目的也是想了解荷兰建筑师的想法。

卡斯：这是不一样的，这里有很多建筑师，我的方法与其他建筑师有很大的不同。它不同于 UN Studio，存在很多的不同。

李：在过去的 70 年里，大家养成了一种盲目追随的习惯，这是一个不好的现象。我认为原因是中国的建筑没有足够好的基础，从 20 世纪 20 年代到 70 年代，这很令人困惑。自从 80 年代中国改革开放以来，大量的外国建筑涌入，人们感到困惑。与荷兰相比，这里有大礼堂，一个 50 年前建成的大型会议中心，即使是现在，其建造水平依然是相当高的。

卡斯：实际上，OMA 的建筑与此联系紧密，但它通常是非常具有追溯性的，它看起来像 50 年代的，也使用了当时的技术。

李：我想是的，它代表了当时高水平的建筑和建筑设计。我认为在荷兰，建筑的老传统很好。

that is the new craft. It is a craft because it relates directly to the material, you can see the design, the production, it relates immediately to the material. And you don't have to look abroad from China, I mean, there's exactly something you can do from inside. That's not something you can import, it's something you have to internalize. So you have to have empathy, and have digital design methods. Then it's not a matter of importing it from abroad, it's just an interior thing.

LI: Yes, you're right, the purpose of my visit here is to learn about how Dutch architects think.

KAS: That is different, there are so many different architects, so my approach is really different from other architects, it's different from UN Studio, even, so there are many differences.

LI: For the past 70 years, it has formed a habit of following, it's a bad habit. I also think the reason is there's not enough good foundations in architecture in China, from 20s to 70s, it's very confusing. Ever since 80s when China opened up to the world, there's been a huge influx of foreign architecture, people are confused. Compared to the Netherlands, there's Aula, a big conference center which was built 50 years ago, and even now it's still a high level of construction.

KAS: Actually architecture of OMA is very much related to that, but it's typically very retroactive, it looks like the 50s, and it also uses technology of that time.

LI: I think so, it represented the high level of architectural design and construction at that time. So I think in the Netherlands, the old tradition of architecture is good.

卡斯：我认为这是一个更大的国际运动的一部分，不仅仅在荷兰，而是一个叫作"粗野主义"的国际运动，与简单的叠加相比，它是相当大胆的。这种大胆的姿态就在那里，在其他国家也有许多很好的例子。在俄罗斯，也有同样的运动，与荷兰的自由主义政体相比，这些表达是相似的。这很有趣，跨越了所有的边界和政治体系。

李：我认为学习如何思考，如何创造新事物和创新是很重要的。

卡斯：我必须说一句。如果我和某人交谈，这不仅是一方从另一方学习，另一方也从前者学到东西。如果我和我的博士霍姆对话，我也能从他那里学到东西，而不是我告诉他去做什么。因此，如果你说学习怎样思考，我宁愿说它是建立对话，而不是输入一种思考方式。我相信这对于我们该如何做至关重要。所有存在的东西都来自于对话，至少是好的事物。坏事来自于隔绝。

李：我的问题是关于中国建筑的，我们的建筑在整个世界所处的位置以及我们应该去向何方。这很重要，也许这对当地人来说很自然，但对我们来说这是一个非常重要的经历。正如你在书中所说，"随着世界不断发展，我们将不得不寻找新的建筑基础"。我可以推断，实现这个目标是从旧元素中找到一个新的

KAS: I think it's part of an international movement, not just in Dutch, but an international movement called "brutalism", and it's quite bold, compared to simple stacking. This boldness of gesture is in there, but there are many other great examples in other countries of brutalism. In Russia, there are examples of the same movement, which is funny because the expressions are similar compared to the liberal regime in Holland. This is interesting, crossing all borders and political systems.

LI: I think it is important for us to learn how to think, how to create something new and innovate.

KAS: I must make one remark. If I talk to someone, this one learning from the other, but the other also learns from the one. So if I talk to Holm, my PhD, I learn from him as well, it's not like I tell him what to do. So if you say learning to think, I'd rather say it is setting up a dialogue, rather than importing away of thinking, I believe this is essential for how we do it. Everything that exists comes from dialogue, at least the good things, the bad things come from separation.

LI: My question is about Chinese architecture, where we are in the whole world of architecture and where we should go. This is important, maybe it's very natural to locals, but to us it's a very important experience. Just as you have said in your book, "as the world keeps turning, we will have to find new foundations for architecture". I can infer that implementation is finding a new order from old elements, and form

秩序，而形式和空间只是一个结果。这个过程更加重要，所以我们应该深入研究。我们不应该只关注形式和空间。如今，建筑的内部元素比几十年前更加重要，比如计算机技术和互联网技术。它完全改变了设计的方式。你认为新秩序是从现代建筑理论中衍生出来的，就像一棵新树从老树的枝杈上长出来的，还是完全独立的？你能阐述一下吗？

卡斯：对我来说，很明显这是进化。大规模生产是为更多人创造房子的进化过程中的一个重要步骤，在中国，它仍在发生，尤其当人口涌入城市，需要一所房子时，这种现象比以往更加显著。在我看来，这是一件好事，但我们知道，建造工厂来生产汽车或建筑部件，预制是很重要的。它在 20 世纪 60 年代得到完善，实际上它有好的社会效益，因为最后人们有了房子。在东欧、俄罗斯、美国也发生过同样的情况，大量的房屋基于大规模生产建造，所以我不批评这一点。但现在我们要做的是多样化，关注每个人及其偏好。开发出来的工具，比如机器人，就是用这种能够使生产多样化的方式发展起来的。因此，从大规模生产开始，这是大规模定制过程中一个非常小的步骤，我们不应该完全排斥机器，而是和它们一起工作，用它们来促进大规模定制。然后，你可以生产许

and space is just a result. The process is more important, so we should research into deeper reasons. We shouldn't just follow the form and space. Today, there are internal factors for buildings which are weighted more heavily than decades before, such as computer tech and internet tech. They changed the way of design completely. Do you think the new order is derived from modern architecture theory, like a new tree stemming from the branch of the old tree, or it is completely separate? Can you expand on it?

KAS: For me it's very clear that it's evolution. Mass production is a vital step in evolution to produce many houses for many people, and in China it's still happening, especially since more than ever, people come to cities and need a house. In my opinion, that is a good thing, but we know, we build factories to produce components for buildings or cars, every prefabrication is important. It was perfected in the 60s, and actually it was a good thing socially, because in the end people had houses. And the same happened in eastern Europe, the same happened in Russia, the same happened in United states, massive houses based on mass production, so I don't criticize that. But now we're looking to diversify, looking at the individual person and their preferences. The tools that are developed, like robots, they were developed in such a way that they are actually able to diversify production. So from mass production, it's a very small step to mass customization, we should not turn our backs to the machines at all, but rather work with them, and use them to

HESSING

多房子，并同时满足一些富人阶层的个人偏好，这也会发生在中国，而不仅仅是富人阶层。我的意思是，首先总是富人，但不幸的是，中国也有富人阶层，到处都是，在美国也一样。这是一个全球性的发展，但从长远来看我们不会接受它，我们希望每个人都有这样的定制，这个想法将会被主流所吸收。这是我的观点，我认为它正在发生。当我看到混凝土工厂里的人，他们习惯于生产，首要任务是客户，他们想要解决单个客户的问题，个别客户的房子、汽车……这是个人偏好，我认为这种情况正在发生，我确信在这里和在中国并没有很大的区别，只是中国的规模如此之大，有如此多的人能让它变得有趣。事实上，我们可以从中学习，你如何与许多人打交道，这就是我们的兴趣所在。在我看来，在设计中如何满足众多个体的需求，才是最重要的事情，因为我们是设计师。

李：另一个问题是在 A 和 B 之间，你倾向 A 还是 B？一棵新树还是一根新树枝？

卡斯：这棵树将自我更新，树有生命。这就像鸡和蛋的问题一样，一棵树和一粒种子也是同样。它们有着非常密切的关系，但对我来说，看到鸡和蛋是一回事，这没什么不同。当然，一棵树生长，然后它结束，但留下新的生命。这是一个循环，但它是一个系统，因此我不能在 A 和 B 之间进行选择。

facilitate mass customization. Then you can produce many houses and at the same time address individual rich's references simultaneously, and that will also happen in China, not only for the rich. I mean, first for the rich, always, and unfortunately in China there's also an upper layer of rich people, it happens everywhere, same here, same in the USA. This is a global development, but we won't accept that in the long run, we want this customization for everyone, that idea will land and be absorbed by the mainstream. That's my view and I see that happening. When I see a presentation of these guys - concrete factories, who are used to producing, on the top of their priority list is customization, they want to address individual customer, individual buyer of house, individual buyer of cars, this is personal preference. That is happening, and I'm sure there's not a big difference between here and in China, only the scale in China is so huge, there are so many people that make it interesting. Actually we can learn from that, how you deal with many people, and that is what interests us here. To find how to deal with flocks of individuals in the design, is ultimately the most important thing, in my opinion, because we are designers.

LI: Another question is between A or B, would you prefer A or B? A new tree or a new branch?

KAS: The tree will renew itself, a tree has a life. But this is like a chicken and egg problem, same thing like a tree and a seed. They have a very close relationship, but to me, it's important to see the chicken and the egg as the same thing, it's nothing different. Of course, a tree grows, and then it ends but gives birth to new life. That's a sort of cycle, but it's one system, so I can't choose between A and B.

李：如此看来，现代建筑就像一块巨大的土地，可以把它发展得很长远。虽然这个时代充满了革命，但它们仍然有着密切的关系。

卡斯：但是你得给它时间。如果你强迫它，压缩时间，虽然改变是好的，但是时间太短，它不再是自然的，因此它会造成问题。你不能在这么短的时间内强迫某件事的发展，它需要时间。

李：我的另一个问题是关于公共原则和个性，我认为前者是基本的。40 年前，所有材料基本上是一样的，而今天，科技在各方面的进步，使建筑师的个性得以彰显。另一方面，这些不同，也来自于我们自身的不同。教授的建筑是流线型的，扎哈的建筑也是流线型的，但两者结构体系是完全不同的。另一个理论是"一数一世界"，这非常有趣，非常方便建造。第三个是如何评价建筑，第一种条件是他们是否在设计中抓住关键问题。其次是他们是否使用了有效的方法来解决问题，第三个条件是它是否给了人们一种清新的感觉。你如何评估一个建筑是好是坏？

卡斯：在这方面我们有一些辅助的东西，但我认为这更多的是在人类层面上。我认为人们可以欣赏 17 世纪的宫殿，可以欣赏我的建筑，可以欣赏哈迪德的建筑，他们可以欣赏许多不同的东西。值得一提的是，欣赏不仅仅是在建筑层

LI: I can sort of see, modern architecture is like a big plot of land, you can develop it very far. Although this age is full of revolutions, they still have a close relationship.

KAS: But you have to give it time. If you want to force it, to compress time, the change was good but the time was too short, and it's no longer natural, therefore it caused problems. You can not force something in so short time, it needs time to evolve.

LI: Another question I have is about common's principle, and another is personality, the former is fundamental, I think. I think 40 years ago, all materials were basically the same, but today, with the progress of science and technology in all aspects, the personality of architects can be highlighted. At the same time, these personalities come from our own differences also. Professor's building is fluent, Zaha's building is also fluent, but the structural system is completely different. Another theory is one building, one detail, it's very interesting, very convenient for construction. The third one is how to evaluate a building, the first condition is whether they catch the key problem in the design. Second is whether they used an efficient way to solve it, and the third is whether it gave people a refreshed feeling . How do you evaluate if a building is good, so so, or bad?

KAS: We have an assistant for that but I think it's more on a human level. I think people can appreciate a palace from the 17th century, they can appreciate one of my buildings, they can appreciate Hadid's buildings, they appreciate many different

面上。举个例子，如果你把一所大学和一个监狱进行比较，它们是非常不同的，因为在监狱里，门是锁着的，但在大学，人们希望门是开着的。尽管我觉得他们被锁得太多了，但这里你也有一个安全系统，就像在监狱里一样。它更多的是关于我们的社会，它是如何组织的，它是否允许人群的流动，它是否允许灵活的、多模式的空间使用。这与建筑没有太大关系，因为人们很有创意，他们可以使用一个旧的工业工厂，只要他们能进去，就可以合理利用它。在建筑方面，如果建筑有特色，那么，就像旧宫殿的特点是坚固一样，对我们的建筑来说也是一样的，这很有帮助，因为你和它有更密切的联系。旧的工业建筑可以很有特色，但你不会去一个现代化的工业区，因为它无趣，没有特色。你是怎么做的，特征塑造？这意味着，你有某种形式的表达和个性，这也是为什么定制建筑比大规模生产的建筑有更多的特征。你不会去一个高层住宅项目说"这太好了，它有特色"。这些建筑将在一段时间后被拆除，50年、30年之后，人们把这些建筑推倒。但那些有特色的建筑，它们还屹立着，我们想保留它们。这就是为什么人们喜欢这些建筑，因为它有特色。你同意吗？

李：所有的特色都值得被欣赏。

things. One thing is that appreciation is not only on an architectural level. For example if you compare a university with a prison, it's very different because in prison, the door is locked, but in another, hopefully the door is open. Although I think they're locked too much anyway, here you also have a security system, like in prisons. It's much more about our society, how it's organized, and does it allow for the flow of people, whether it allows for flexible, multimodal use of spaces. It's not so much to do with architecture, because people are very inventive, they can use an old industrial factory, and use it reasonably as long as they can go into it. In the architectural part is if the building has character, so the character of the old palace is strong, same for our buildings, and that helps because you have a more personal bond with it. An old industrial building can have a nice character, but you won't go to a modern industrial area, because it's not interesting, it has no character. How do you do that, character building? It means, you have some form of expression and individuality, that's also why customized buildings have more character than mass produced buildings. You won't go to a high-rise housing scheme, and say "this is great, it has character", not really. So those buildings will be taken down after a while, 50 years, 30 years, they take these buildings down. But the buildings with character, they stand, we want to keep them. That's why people like this building, because it has character. Do you agree?

LI: All characters are worthy to be appreciated.

卡斯：是的，但是有特色和没有特色之间有什么区别呢？这是一种融合，建筑占了一部分。我不认为建筑在这里扮演着最重要的角色，但它也有一定的作用。

李：我接受过传统教育。许多人喜欢问什么是标准。因为在中国更高的标准是好的，而更低的是不好的。另一个问题是，你的书（《Towards a New Kind of Building》）中提出了一个观点，叫作"一数一世界"，这个观点对我来说非常重要。我在互联网上看过你的作品，它流畅又有力，我很欣赏它，但我不知道怎么去构建它。当我看到你的作品时，我认为它反映了很多社会责任感，不仅是一个好的建筑，而且是对社会有益的东西。因为它的观点可以使建筑变得非常简单，这与扎哈完全不同。我还有一个关于美国建筑史的问题，你的观点与富勒的理论有什么联系吗？

卡斯：我想是的，我读过他的书，他是一个发明家。

李：他们也使用相同的颜色作为"建造大空间"，我认为这不仅是建筑，也是工程。它涉及很多其他的方面，我认为你的也是，我真的很喜欢它。我也有这本书（《Towards a New Kind of Building》），不知道它是否被翻译成了中文。虽然我刚读了第一部分，但我认为你的观点对中国建筑师非常有用。

KAS: Yes, but then what is the difference between character and not having character? It's a mix, part of it is architectural. So I don't think architecture plays the most important role here, but it has a role.

LI: I had a traditional education. Many people like to ask what is a standard. Because higher standard is good, and lower is bad in China. Another question is that your book (*Towards a New kind of Building*) offers an opinion called "one building, one detail", it's very important to me. I think on the internet, I've seen your work, it's very fluent, strong, and I admire it, but I don't know how to construct it. When I saw your work, I think it reflected a lot of social responsibility, not just good architecture, but beneficial to society, because it's opinion can make construction very easy. It's quite different from Zaha, and I also have a question about American architecture history, does your opinion have a relationship with Fuller's theories?

KAS: Yes, I think so, I've read his books, he's an inventor.

LI: They also use same colors as "build great space", and I think it's not only architecture, but also engineering. It's multiple aspects are others, and I think yours is also, I really like it. I also have a book (*Towards a New kind of Building*), I don't know if it's translated to Chinese. Although I just read the first part, I think your opinions are very useful to Chinese architects.

卡斯：出版中文版，这是个有趣的想法。

李：这不仅仅是一本互动的建筑书，里面也包含一些非常基本的原则（很有个性）。

卡斯：扎哈的作品和我的作品之间的区别其实很简单，扎哈更多的是外部想法，关注从外部看到的，我想要做的是从内部、从组件，以及如何把它们组织到一起，但是，这种关系仍然与外部因素和环境有关。因此它完全基于一个理论，它是一个动态的系统。无论我们做什么，它都是一个动态系统。动态是内在的，这些元素一起作用，在一个动态的环境中，这是所有事物的基础。

李：相反，中国的一些年轻建筑师可以设计出特殊的形式和空间，但如何科学地建造是一个重要的问题。

卡斯：一个例子是中国北方的某个剧院。它有一个外部形态，但如果你看到它是怎么做的，那就没有意义了。

李：是的，这在空间上很浪费。很多材料放进去，在中国是一种常见的大趋势。我认为好的建筑也应该有好的建造。建筑师应该知道如何使用材料。这就是我所说的中国的混乱，我认为这是很不负责任的。

KAS: It's an interesting idea, to make a Chinese version.

LI: It's not only an interactive building book, but it also has a very basic principle (It has character).

KAS: The difference between Zaha and our work is actually quite simple, Zaha is much more about exterior idea, what's seen from the outside, what I try to do is approaching from the inside, from the components, and how they organize themselves together, but still, the relationship is with external components, with context that has influence. So it's fully based on an idea that it's a dynamical system. Whatever we do, it's a dynamical system. The dynamics are inside, and those elements act together, but they are also in a dynamic environment, that's really at the basis of everything.

LI: On the contrary, some young architects in China have special form and space, but how to construct scientifically is still an important question.

KAS: A example is a theater in North of China. It has an external shape, but if you see how it's made, it doesn't make sense.

LI: Yes, it's very wasteful in space. Lots of materials were just put in, it's a formal, big trend in China. I think good architecture should also have good engineering. It should know how to use material. So that's why I'd like to tell you there's confusion in China. I think it's very irresponsible.

卡斯：但在荷兰，这个行业也很分散。从一个阶段到另一个阶段，需要有设计师、工程师、顾问等。然后会有文件，一切都是固定的，接下来是承包商，等等。承包商必须找到制造商，但是承包商并不真正理解设计，会出现很多混乱和大量的能量浪费，我说的是荷兰的情况。最终，他们必须要把它组装起来，让它看起来像设计师设计的样子。但是这个过程没有从制造商到设计师的反馈，我们希望设计深入到生产过程，然后再反馈回来。我们想要这两个端点，机器，以及这些机器如何运作，我们希望它们与设计过程相连。我们不希望它只是一个线性过程，而是所有的参与者都积极参与其中。在中国，这样做更难。

李：在这里我发现一个有趣的现象，它与我的下一个问题有关，关于艺术和技术。我认为荷兰的建筑是理性的，有很多水平的线条，也许是因为它很容易造，并且改变了材料、颜色，让它看起来很好看。这很合理，代表着一种思考方式。荷兰有很多水平线条，这也许和建筑方法有关，或者和这里的地形有关，大多是平坦的，没有山。

KAS: But also in Holland the industry is very fragmented. You go from one phase to the other, so you have the designer, the engineer, consultants etc. Then you have documents, everything is fixed, then you go to the contractor and so on. The contractor has to find manufacturers, but then the contractors don't really understand the designs, there's a lot of confusion and a lot of energy is lost, I'm talking about the Dutch situation. In the end, they somehow have to make it and put it together, and it has to look like what the designer designed, but there is no feedback from the manufacturer to the designer, we want the design to production process, and back again. We want those two extremes, the machines, and how those machines talk, we want them to be connected to the design process. We don't want it to be just a linear process, but all those players be active together. It's even harder to do that in China.

LI: I have found an interesting phenomenon here, and it's related to my next question, about art and tech. I think Netherlands architecture is rational, with a lot of horizontal lines, maybe because it's easy to construct, and changes the material, the color, to make it good looking. It's very reasonable, representing a way of thinking. So there are so many horizontal lines in Netherlands, maybe it has to do with construction methods, or the landscape, which is horizontal, no mountains.

卡斯：在我看来，荷兰是一个三角洲，河流汇集在一起，一切都变得平缓，这是一切都来到这里都变得平坦的想法的一个条件。这也是为什么在政治中，你可以用隐喻来表达，但他们寻求妥协，把它压平。因此这个国家地理特征与人们的行为有某些关联，我认为情况是这样的。但从另一方面来看，它一直在变化，因为在荷兰，政治体制已经行不通了，这里也有混乱。所以我们可以有理论，但它不是固定的，它一直在变化。

李：与德国和法国相比，这里有许多水平线条。法国是浪漫的，而荷兰更加理性。

卡斯：你也可以从语言中注意到，荷兰语很单调，法语很好，我喜欢法语。

李：我认为技术是一个基本要素，但要如何组织出更好的艺术形式呢。但是在中国，这方面退步了，我不关心技术，我认为这是一个不好的习惯。技术不仅与建筑有关系，它也是美学的一个重要的基础，因为技术可以成为艺术的源泉，而不仅仅是艺术。它不同于纯粹的艺术，比如绘画，它不需要很多技术手段。但建筑艺术更依赖于技术。

KAS: The way I see it is Holland is a delta, the rivers come here together, and everything flattens out, the idea that everything comes here and flattening out is a condition. That's also why in politics, you can speak in metaphors, but they look for compromise, flattening it out. So there's something in the character of the country geographically that is related to what people do. I think that is the case. But on the other hand, it changes all the time, because in Holland the political system doesn't work anymore, there's also confusion here. So we can have theories, but it's not fixed, it changes all the time.

LI: Compared to Germany and France, there are so many horizontal lines. France is very romantic, the Netherlands is much more rational.

KAS: You can also notice from the language, dutch is very monotone, in French, it's very nice, I like the French language.

LI: I think technology is a basic element, but how to organize better to form art. But in China there's recession, I don't care about technology, I think it forms a bad habit. Technology not just has a relationship with construction, but it's also a great foundation of aesthetics, because technology can be the source of art, and not just art. Because it's different from pure art like drawings, which doesn't need a lot of technology. But architecture art has more reliance on technology.

卡斯： 作为设计师，更多的是处理各种关系。它不再仅仅是一个草图的距离。不是作为科学家，因为他实际上是在使用新的数字技术来实际建造东西，所以通过这样做，不是通过谈论它，不仅仅是思考，不仅仅是在材料上，它会变得更像是材料本身，并重新定义它。我认为这样看它是有趣的，然后科学家使得自己越来越像设计师，这又让几乎每一个人都是设计师，因为通过你的图像，Photoshop 的工作人员可以做更多的图像出来，也有程序可以改变你的脸，等等。每个人在这个设计的过程中都变得越来越完善，我认为所有的因素被放在一起，这是很有趣的。这也是国际的，这是进化。

李： 是的，你从基本的元素，比如材料中思考一些东西，而这种材料决定产生什么样的外观，这是合理的。另一个我感兴趣的事情是，在两个月前，我来到中国驻荷兰大使馆，我穿过了一个古老的森林，那里面只有一条小路，森林的其他区域都没有受到影响，因为人们与自然保持距离，并未去打扰它。我认为这是一种看待人和自然关系的好方法，我想这可能和人们的信仰有关系。

KAS: As a designer, it's more about dealing with all kinds of relationships, it's no longer just at a distance of sketch. Not for the scientists, because he is actually working with the new digital technology to actually construct things, so by doing, it's not by talking about it, not only thinking, not only in material, it becomes more working the material itself, and redesign it. I think it is interesting by looking at it, then scientists make themselves more and more like designers, and this again, makes almost everyone a designer, and because it can do a lot of Photoshop staff already with your images, also there is program that you can change your face and etc. Everyone becomes more and more improve in process that could be called a design process, I think that everything is being put together and it is interesting. That's also international, this is evolution.

LI: Yes, you think about something from the basic element like material, and this material can result what kind of the looking, this is reasonable. And another thing that I'm interested in is that one time, in two month ago, I have come to the Chinese embassy in The Hague, I have walked around through an old forest, there is just a little way in it, and all the forest in the area is not affected, because people keep a distance with the nature, do not disturb it. I think it is a great way to look at the relationship between people and nature, I think it may have a relationship with people's believing.

卡斯：你知道吗，荷兰所有的森林都是种植出来的，而不是自然森林，所以一切都是新的种植，当然有些已经很老了，一切都是新植物。你给出的例子也是这些树的开始，我的意思是它不是一个天然的森林或者什么东西，但是你可以在 50 年内创造出森林，但这是一个相对的概念。你可以在 50 年内在亚马孙建造一片森林，在 50 年后，它就有 90% 的容量了。它的生成速度相当快，但它必须被种植，它不会自己开始，这是人们所做的。它没有在很多地方发生，因为我们需要更多的环保人员。

李：最后，我想写一本关于建筑创新过程以及最初的观点和方法的书，显然想要提到教授您的理论，你能给我一些建议吗？

卡斯：怎样写这本书？

李：是的，我想写一本对国内建筑师有用的书，我认为这是一个开始。这里有一件趣事，大约在 2002 年，有一个关于广州大剧院的投标，有很多著名设计师参与竞标，包括扎哈和库哈斯。我认为库哈斯的方案很好，应该拿下那次投标，但政府官员认为它不够有力，他们喜欢扎哈的设计，所以扎哈的设计被选中了。但是库哈斯赢得了位于北京的央视大楼的竞标，这个项目不是很好，但是在那里，他正是给了人们真正想要的东西。

KAS: You know, all the forest in Holland are planted, not nature forest, so everything is new planted, of course some are already old, everything is new plant. The example you give is also a start of these trees, I mean it's not a natural forest or something at all, but you can create forests in fifty years, but it's a relative concept. You can build jungle in fifty years, it has ninety percent of its capacity. It is quite fast in generation, but it has been planted, it didn't start by itself, that's what people did, not happened everywhere because we need more this green staff.

LI: At last, if I want to write a book about the innovation procedure, and the original point and the means and the way obviously want to include professor's theory, so could you give me some advise?

KAS: How to write the book ?

LI: Yes, I want to write the book, it is very useful to Chinese. I think it is a beginning. About 2002, there is a competition about Guangzhou Theatre, a lot of special designers come here including Zaha and Koolhaas. I think Koolhaas's job is good, which should get the nod, but officers think it is not strong enough, they like Zaha's job, so Zaha have been chosen. But he wins the Beijing CCTV competition as you know, this project is not good, but in there, I give you just what you really want.

卡斯：我不相信那一点，那是个谎言。事实上，在库哈斯的作品中有非常多美好的东西，只有贴近观察才是深入理解这些作品的方法。关于CCTV大楼，如果真如你说的那样，我认为这真的很糟糕。他应该发挥创造力，因为他想要，他和他的团队想要，而没有人想要。他选择这个模式不是因为别人想要它，而是因为他想要它。当然其他人也很欣赏它。有人可能会说没关系，我们来做吧，但我不认为它特别好，我认为它背后的理论真的很薄弱，理论上可能有一件好事但实际上它不起作用，因为人们说我想要一个循环，人们可以继续，但是没有人继续在大楼里，所以这个理论与他所做的是不同的，而且它也花费了很多钱，我认为这是最大的失败，如此昂贵，你甚至不应该计划这么做。那是历史，他设计了一件很贵的东西，他知道这一点，他对中国人说，你们就按照我说的做，他责怪你们是因为你们按他的建议去做，我不喜欢那样。

李：我不认为这个项目很好，它没有遵循基本原则，它的顶部结构非常重，这只会造成巨大的浪费。

KAS: I don't believe that, that is a lie. Actually, in all Koolhaas' works, there are a lot of pretty and this is the way to actually close himself for that pretty. About CCTV, If it is indeed, I think it is really bad. He should take the creative because he wants it and no one else want it, just he and his team. He chooses the model not because someone else want it, but because he wants it. And of course other people appreciate it somehow. Someone may say it's OK, let's do it, but I don't think it is specially good, I think the theory behind it is really weak, there is maybe one good thing in theory but in practice it doesn't work, because people said I want a loop that people can continue, but no one continues inside the building. So the theory is separated from what he is doing and it also costs a lot of money, I think that is the biggest failure, so expensive, you should not even plan to do that. That is histories, he designs something that is so expensive, he knows that and he says to Chinese, you just do what I say, and he blames you because you do what he proposed, I don't like that.

LI: I don't think this project is a good one, it's not following the basic principle, the structure is very heavy on the top, it's just a huge waste.

卡斯：当然，这对很多人来说是一个挑战，我们必须解决它。但如果你看基础，你必须有巨大的柱子，我的意思是需要在里面加很多钢材。假如你知道你正在做什么，这真的很惊人。所以它的力并不集中，我想，他用错了力，我不喜欢这样。

李：是的，我也不喜欢这样，这并不好。

卡斯：这不是中国人的兴趣所在，这太奇怪了，这是不负责任的，这是很困难的。

李：我认为我在书中所写的基本原因是责任。我认为这是一个很好的源头，很好的方式引导了一个很好的结果，它应该是明确的标准，这是我想写书的原因。

KAS: Of course it is a challenge for many people, we have to solve it. But if you look at the base, you have to form huge column, I mean so much steel in that. If you know what you could have done, it's amazing. So it's informal concentration of power, I think through it, he misused that power, so I don't like that.

LI: Yes, I also don't like that, it's not a good one.

KAS: It's not in the interest of Chinese people, it is so weird, it is irresponsible, this is difficult.

LI: I think the basic reason that I write in my book is the responsibility. I think it is a good original and good way leading to a good result, it should be clear standard here, and it's the reason I want to write a book.

图片来源

P07 代尔夫特理工大学大礼堂（TU Delft Aula）华南理工大学建筑学院　李晋

代尔夫特理工大学图书馆（TU Delft Library）华南理工大学建筑学院　李晋

P09 《印象·雾（小港）》莫奈，现藏于巴黎马默顿博物馆

P10 《埃斯泰克的海湾》塞尚，现藏于芝加哥艺术学院

《静物苹果篮子》塞尚，现藏于芝加哥艺术学院

《圣·维克多山》塞尚，马迪拉藏

P13 乌特勒支大学学习中心　华南理工大学建筑学院　李晋

P18 阿姆斯特丹单亲家庭住宅　华南理工大学建筑学院　李晋

阿姆斯特丹孤儿院　华南理工大学建筑学院　李晋

海牙小天主教堂　华南理工大学建筑学院　李晋

P27 图 1-01　卡恩建筑事务所（KAAN Architecten）摄影师 Inga Powilleit

P29 图 1-02　卡恩建筑事务所 摄影师 Fernando Guerra FG+SG

P31 图 1-03　卡恩建筑事务所 摄影师 Fernando Guerra FG+SG

图 1-04　卡恩建筑事务所 摄影师 Fernando Guerra FG+SG

P33 图 1-05　卡恩建筑事务所

图 1-06　卡恩建筑事务所 摄影师 Fernando Guerra FG+SG

P35 图 1-07　卡恩建筑事务所

图 1-08　卡恩建筑事务所

图 1-09　卡恩建筑事务所 摄影师 Fernando Guerra FG+SG

P37 图 1-10　华南理工大学建筑学院　李晋

图 1-11　卡恩建筑事务所 摄影师 Fernando Guerra FG+SG

P39 图 1-12　卡恩建筑事务所 摄影师 Fernando Guerra FG+SG

图 1-13　卡恩建筑事务所 摄影师 Fernando Guerra FG+SG

P41 图 1-14　卡恩建筑事务所 摄影师 Fernando Guerra FG+SG

图 1-15　卡恩建筑事务所 摄影师 Fernando Guerra FG+SG

P43 图 1-16　卡恩建筑事务所 摄影师 Fernando Guerra FG+SG

P45 图 1-17　卡恩建筑事务所 摄影师 Beauty & The Bit

P47 图 1-18　卡恩建筑事务所 摄影师 Beauty & The Bit

图 1-19　卡恩建筑事务所 摄影师 Beauty & The Bit

图 1-20　卡恩建筑事务所 摄影师 Beauty & The Bit

P49 图 1-21　卡恩建筑事务所 摄影师 Beauty & The Bit

图 1-22　卡恩建筑事务所 摄影师 Beauty & The Bit

P51 图 1-23　卡恩建筑事务所 摄影师 Fernando Guerra FG+SG

P53 图 1-24　卡恩建筑事务所 摄影师 Fernando Guerra FG+SG

P55 图 1-25　卡恩建筑事务所

P57 图 1-26　卡恩建筑事务所

图 1-27　卡恩建筑事务所 摄影师 Fernando Guerra FG+SG

P59 图 1-28　卡恩建筑事务所 摄影师 Fernando Guerra FG+SG

参考文献

[1]   Paul Groenendijk & Piet Vollaard. 2006. *Architectuurgids Nederland (1900–2000)* [M]. Rotterdam: Uitgeverij 010 Publishers.

[2]   Paul Groenendijk & Piet Vollaard. 2009. *Architectuurgids Nederland (1980–nu)* [M]. Rotterdam: Uitgeverij 010 Publishers.

[3]   Jurgen Joedicke. 1963. *Dokumente der modernen architektur herausgegeben von jürgen joedicke uitgeverij g van saane lectura architectonica hilversum* [M]. Stuttgart: Karl Kramer Verlag.

[4]   Belinda Thomson. 2010. *Van Gogh GEMÄLDE* [M]. Amsterdam: Van Gogh Museum / Mercatorfonds.

[5]   Robert McCarter. 2015. *Aldo van Eyck*. Robert McCarter.

[6]   Atelier Bow-Wow. 2010. *Behaviorology* [M]. New York: Rizzoli.

[7]   Paul Groenendijk, Piet Vollaard Rook & Nagelkerke. 2010. *Architectuurgids Rotterdam* [M]. Rotterdam: 010 Publishers.

[8]   Francine Houben & Herbert Wright. 2015. *Mecanoo architecten: People, Place, Purpose* [M]. Artifice books on architecture.

[9]   Prestel Verlag. 2011. *Rembrandt* [M]. Munich, New York: Prestel Publishing Ltd.

[10]   Francine Houben & mecanoo architecten. 2008. *Dutch mountains*. Uit everij de Kunst.

[11]   Juan José Lahuert & Pere Vivas. 2014. *CASA BATLLÓ*. Ricard Pla.

[12]   Aukje Vergeest , 李梅 . 2015. 直面——文特森·凡·高 [M]. Amsterdam: Van Gogh Museum.

[13]   Robert McCarter. 2006. *Herman Hertzberger* [M]. Rotterdam: Nai 010 publishers.

[14]   Muriel Grindrod. 1972. *World Cultural Guides* [M]. Schaan: Park and Roche Establishment.

[15]   Rem Koolhaas. 1996. *Delirious New York* [M]. New York: The Monacelli Press.

[16]   Onder redactie van José ten Berge-de Fraiture en Annemarie van Oorschot-Dracht. 2015. *Pastoor Van Ars-Monument van Aldo van Eyck*. Ars Architectuur Comité.

[17]   Kas Oosterhuis. 2011. *Towards a New Kind of Building* [M]. Rotterdam: NAi Publishers.

[18]   Toyo Ito, Kumiko Inui, Sou Fujimoto, Akihisa Hirata & Naoya Hatakeyama. 2013. *Architecture. Possible here? "Home-for-All"* [M]. Tokyo: TOTO Publishing LTD.

[19]   Peter Zumthor. 2014. *Thinking Architecture* [M]. Basel, Boston, Berlin: Birkhauser.

[20]   O.M.A. Rem Koolhaas and Bruce Mau. 1995. *S,M,L,XL* [M]. New York: The Monacelli Press.

[21]   Liane Lefaivre and Alexander Tzonis. 1999. *Aldo van Eyck Humanist Rebel* [M]. Rotterdam: 010 Publishers.

后　记

当代的荷兰建筑现象纷繁多元，如果从其外在表象观察荷兰建筑往往会产生无所适从的茫然。正如苏轼的《题西林壁》中所描绘的庐山："横看成岭侧成峰，远近高低各不同"的眼花缭乱。而若转换视角，从产生这些表象之根源来观察荷兰建筑，则一切会顺理成章。一方面，这种纷繁多元展示了不同的设计主体对于建筑不同角度的阐释，而这一切源于对于感知的尊重，不论感知来自于个体感受还是群体的认知；另一方面，突飞猛进的荷兰建筑技术为这一切提供了保障。在荷兰建筑中，我们看不到对于现代主义所界定的多种关系（形式与技术、形式与功能等）合理性的纠结，我们看到的是围绕建筑核心问题的多因素交织的顺理成章，以及对建筑边界问题的不断触及与尝试。

本书作为一本对谈录，一方面展示了当代荷兰建筑师一部分中坚力量不同的态度与方法；另一方面，我们也可以提炼出他们思考方式的共同本底，而后者更值得我国建筑师借鉴。从完成访谈到本书的梳理完成经过了三年多的时间，在这段时间里，经过了将荷兰建筑"超级现象"本底根源，转化为可视的历史脉络的过程；也经过了自己作为一名老师和建筑师独立的印证过程。我们所看到的荷兰现代建筑现象，有着深厚的历史、文化与技术的根源，而对于根源的探究是有助于我国建筑师更新思维的角度。本书在这方面仅仅做了一点工作，不足之处，请同行与爱好者指正。

为说明某些建筑作品细节引用了来自谷德、ArchiDaily、中国建筑官网的部分图片。

本书的最终出版首先感谢中国建筑工业出版社徐晓飞先生的建议、鼓励与耐心；感谢校友冯翰博士、林颖豪先生、杨森先生在荷兰的帮助，为访谈创造了环境与条件；感谢同事朱亦民老师、周毅刚老师对本书提出的建设性意见；以及多名研究生对书的出版所提供的帮助。

感谢 Kees Kaan，Jacob van Rijs，Nathalie de Vries，Francine Houben，Kas Oosterhuis 五位建筑大师对自身设计理念的分享，使自己受益良多。

谨以此书献给我的父亲母亲、妻子孩儿。

二〇二〇年秋日　于华园

图书在版编目（CIP）数据

再认识 = Rethinking Modern Dutch Architecture
: Dialogue with Contemporary Dutch Architects : 荷
兰现代建筑对谈录 / 李晋编著 . -- 北京：中国建筑工
业出版社，2021.9
 ISBN 978-7-112-26437-7

Ⅰ.①再… Ⅱ.①李… Ⅲ.①建筑艺术—研究—荷兰
Ⅳ.① TU-865.63

中国版本图书馆 CIP 数据核字 (2021) 第 154909 号

责任编辑：徐晓飞　张　明
责任校对：王　烨

再认识　荷兰现代建筑对谈录
Rethinking Modern Dutch Architecture: Dialogue with
Contemporary Dutch Architects
李晋　编著
*
中国建筑工业出版社出版、发行 (北京海淀三里河路 9 号)
各地新华书店、建筑书店经销
北京雅昌艺术印刷有限公司制版
北京雅昌艺术印刷有限公司印刷
*
开本：787 毫米 ×1092 毫米　1/16　印张：11 字数：192 千字
2021 年 9 月第一版　2021 年 9 月第一次印刷
定价：78.00 元
ISBN 978-7-112-26437-7
　　(37853)